THE

LOUISIANA SEAFOOD BIBLE

THE LOUISIANA SEAFOOD BIBLE

Shrimp

Jerald and Glenda Horst

PELICAN PUBLISHING COMPANY

GRETNA 2009

First printing, April 2009
Second printing, December 2009

*The word "Pelican" and the depiction of a pelican are trademarks
of Pelican Publishing Company, Inc., and are registered in the
U.S. Patent and Trademark Office.*

Library of Congress Cataloging-in-Publication Data

Horst, Jerald.
 The Louisiana seafood bible : shrimp / Jerald and Glenda Horst.
    p. cm.
 Includes index.
  ISBN 978-1-58980-643-6 (hardcover : alk. paper) 1. Cookery (Shrimp) 2.
Cookery, American—Louisiana style. 3. Shrimps—Louisiana. 4. Shrimp
industry—Louisiana. I. Horst, Glenda. II. Title.
  TX754.S58H67 2009
  641.6'9509763—dc22

                                2008041763

Printed in China

Published by Pelican Publishing Company, Inc.
1000 Burmaster Street, Gretna, Louisiana 70053

To our parents

To Wilbert Horst, who instilled a work ethic in me and showed me how to be a man.

To Hilda Horst, who lit my fires of curiosity and inspired a love for food and innovative cooking that has never left me.

To Cloyd "Nappy" Ray, who indulged my tomboy years by teaching me to fish, catch toads for bait, and play poker for matchsticks, but also insisted I learn to be a lady like my mother.

To Elsie Ray, who taught me the basics of good cooking and a strong love of family.

"There is no love sincerer than the love of food."
George Bernard Shaw (1885-1950),
Irish dramatist and socialist

"Shrimp is the fruit of the sea. You can barbecue it, boil it, broil it, bake it, sauté it. There's, um, shrimp kebabs, shrimp Creole, shrimp gumbo. Pan-fried, deep fried, stir fried. There's pineapple shrimp, lemon shrimp, coconut shrimp, pepper shrimp, shrimp soup, shrimp stew, shrimp salad, shrimp and potatoes, shrimp burger, shrimp sandwich. That—that's that about it."
Mykelti Williamson as "Bubba" in *Forrest Gump*

Contents

Preface

The information in this book is the result of a thirty-year career with the Louisiana seafood industry. As an employee of Louisiana State University Agricultural Center, one of my assignments was to develop better methods of catching, processing, and marketing seafood, from the sea to the table, so to speak. Doing so, I received an intimate view of the industry that few outside of it are privileged to see.

During the course of that time, I also published a monthly fisheries newsletter called *Lagniappe*. Each edition ended with a recipe. Many of those recipes are in this book. Some of the recipes came from our own personal test kitchen, though less than 10 percent of our efforts made the cut and were printed in the newsletter. Many of the recipes were sent in by readers of the newsletter with the understanding that they would be shared with others. Each of these recipes was also tested and about half were deemed worthy of printing. We especially prized recipes that showed originality in ingredients or cooking styles.

Many, if not most, cookbooks are compilations of recipes that sound good when read. To avoid copyright problems, compilers often change one or two ingredients in a "borrowed" recipe. When ingredients are changed in this way, the character of the recipe changes. The recipe seems good, but it hasn't been tested. Because each recipe in this book is tested, we strongly recommend that the first time you try one, you follow it exactly. Tinkering can come in later efforts.

Far more than a cookbook, however, this volume was intended as a food book. As such, it is a complete compendium of information on the subject of Louisiana shrimp that any seafood lover could or would want to know.

This book is a joint effort with my wife, Glenda. It is written in the first person in my voice so that the reader will not have to switch between our voices.

Cook, eat, and enjoy!

Jerald Horst

Acknowledgments

We are grateful to the many people who helped make this book a reality. Marty Bourgeois, Pete Gerica, James M. Nance, Harlon Pearce, Don Schwab, and Tommy Simmons patiently provided advice in response to our many questions. Mark Schexnayder provided both advice and access to important archived material. Thanks are also extended to Lainie Selle and Victoria Zachary for technical assistance. And of course, without the many people who donated recipes over the years the book would not be what it is.

We are indebted to Byron Despaux, commercial shrimper from Barataria, Louisiana, and Don Schwab and Randy Pearce, owners of DoRan Sea-Pak, for welcoming us with our camera into their world of shrimp.

Sincere thanks are extended to the following institutions and individuals for generously sharing their photographs and images: Gerald Adkins, Daniel Alario, Patsy Assevado, Loretta Brehm, Jeanette S. Brown, Karen Chauvin, Kim Chauvin, Harry Cheramie, Malcolm L. Comeaux, Carl Couvillier, Byron Despaux, the Louisiana Seafood Promotion and Marketing Board, the Louisiana Sea Grant College Program, Arthur Matherne, Russ Miget, Michael Moody, the Morgan City Archives, Cecile Robin, Susan Robin, the Texas Parks and Wildlife Department, and May Usannaz.

Special thanks are provided to our manuscript advisory committee: Marie Bezet, Ginger Corkern, Beth Fontenot, Joanne Squires, Tommy Simmons, and Cheramie Sonnier, the latter two of Baton Rouge's *The Advocate*.

Ultimate thanks are reserved for Marie Bezet and Ginger Corkern, who not only served on our advisory committee but labored long and strenuous hours in our kitchen preparing recipes for our food photographer Chris Granger.

The LOUISIANA SEAFOOD BIBLE

Part I: Shrimp

Shrimp:
The New Snob Food?

Americans can be very trendy, especially about food and drink. Holding forth on subtle taste differences in food, especially with jargon-laced language, apparently makes us feel sophisticated. A scant thirty years ago, Americans probably drank as much Tokay wine as Bordeaux. Now we discuss wine with a vocabulary that describes tastes such as honeysuckle, hazelnut, peaches, toffee, pear, steel, oak, grain, coffee, mushrooms, and many, many more. Heady stuff for folk, most of whose ancestors were paupers, peasants, or horse thieves and who arrived on our shores in the dank, dark steerage of sailing vessels and tramp steamers.

Courtesy Louisiana Seafood Promotion and Marketing Board

From William D. Chauvin Collection, Louisiana State Archives, Baton Rouge

Shrimp has now entered that magical circle of foods that has caught the attention of the discriminating gourmet. Until recently, except to those lucky few who lived in coastal fishing communities of the Gulf of Mexico and the South Atlantic, shrimp were, well, just shrimp.

All of that has changed. I recently read food maven Sam Gugino's epistle entitled "Big Time for Shrimp." In it, he said, "When I sucked on the heads of fresh white Louisiana Gulf shrimp . . . , the earthy, primal taste reminded me of an old Burgundy." A few years ago, virtually no one in this country, with the exception of chefs, saw a shrimp with its shell on, let alone one with its head still attached.

Part of the shrimp revolution has to do with improved transportation and communications. The same two thousand species of shrimp in the world that exist today existed twenty years ago. But now, people can access more of them, in more forms, from more places around the world. Well over three hundred different species are commercially important and about seventy are traded worldwide. Almost 90 percent of all the shrimp consumed in the United States is imported and half of that is farm raised.

The commercial shrimp species of the world can be roughly lumped

From William D. Chauvin Collection, Louisiana State Archives, Baton Rouge

into three groups: freshwater shrimp, coldwater shrimp, and tropical or warmwater shrimp. The main commercial species of freshwater shrimp is the giant Malaysian prawn, most of which are now farm raised. A few of these farms even exist in the United States. Other, smaller freshwater shrimp exist. One such shrimp is the river shrimp, a very common species in Louisiana's major rivers.

Coldwater shrimp, as can be expected, are harvested from the much colder waters of the North Pacific and North Atlantic Oceans. Most are small shrimp. The northern shrimp accounts for 80 to 90 percent of the global catches of coldwater shrimp. They range from 90 to 300 count per pound after peeling and cooking. The ocean shrimp of the North Pacific is smaller, ranging between 250 and 300 count after peeling and cooking. As small as coldwater shrimp are, they may need 3 or more years to reach even these sizes. Much smaller numbers of larger, specialty coldwater shrimp, such as the spot shrimp and sidestripe shrimp, are also harvested on a limited basis. Coldwater shrimp are not farm raised.

The last group, tropical shrimp, is where the real action is. A dizzying number of species from North America, Australia, South

From William D. Chauvin Collection, Louisiana State Archives, Baton Rouge

America, and Asia are traded commercially. Eight different species of white shrimp alone can be found among tropical shrimp. They are fast growing and seldom live more than a year. Exceptionally large specimens of larger species can reach a count of two to the pound (eight ounces each). Some species are extensively farm raised, such as the Pacific white shrimp, Chinese white shrimp, and black tiger shrimp. These three species and many more are wild caught as well.

With each species, the taste of shrimp can vary widely depending on the season, their size, how they were caught, or whether they were caught in low-salinity estuaries, the deep briny Gulf of Mexico, or somewhere in between. However, Louisiana is favored by being the primary producer of the queen of the world's shrimp, the Gulf white shrimp—just one of several species caught in Louisiana waters—by consensus the best-tasting shrimp available.

Seafood for Sale:
A Snapshot in Time

Seafood has always been a large part of the New Orleans food marketing scene. The accompanying newspaper advertisement from June 1935 lists lake shrimp for sale at three pounds for ten cents. The lake shrimp mentioned in the advertisement were saltwater shrimp, invariably white shrimp, as opposed to the very common but smaller freshwater river shrimp harvested from the Mississippi River.

The ad also introduces Dominick Natal as the manager of Frey's Fish and Meat Department. Frey's Inc., was owned by L.A. Frey, a name still to be found in the Louisiana food business.

The Frey family name was very prominent in the New Orleans grocery business. In 1888, Nick Frey, a native of Alsace, at that time part of Germany, started a retail grocery store at the corner of New Orleans' Bayou Road and Johnson Street. In 1893, he bought out a grocery at Chartres and Ursulines Streets and some time before 1910, the main store was moved to 1031-1035 Decatur Street.

Nick Frey's son, Frank Frey, succeeded his father and in November 1915 opened a very large store on Canal Street between Camp and St. Charles Streets. The store pioneered the departmentalization concept and was described as "probably the only grocery of its kind in the South" in the May 1925 edition of *The Louisiana Grocer*.

Besides having departments for groceries, fruits and vegetables, meat, poultry, seafood, candy, and a delicatessen, it had a four-ton ice-making machine and refrigerator for fish and game.

Dried Shrimp History

Louisiana has the oldest shrimp fishery in the Gulf of Mexico region. Before ice machines and freezers came into use, Louisiana fishermen harvested shrimp for canning and drying. Louisiana is unique in that it is the only state still harvesting shrimp for drying.

Sun-drying of shrimp was introduced by Chinese immigrants to San Francisco Bay in 1871. One historical account has a Chinese immigrant named Lee Yuen establishing the first drying platform in Barataria Bay in the mid-1860s. Whether or not shrimp drying had been introduced to Louisiana by that time, all accounts agree that by 1873, the newcomers had extended the industry to the bays and estuaries of the state. In 1885, when Yee Foo was issued a patent for the process to sun-dry shrimp, the industry was already firmly established.

Although dried shrimp were first sent from Louisiana to Asian communities on the U.S. Pacific Coast, the state's abundant shrimp harvest soon allowed distribution to Asia, the Philippines, and Hawaii, as well as to the West Indies and South America, though to a lesser extent.

The first drying platform was built in Barataria Bay at a site that came to be called Cabinash. Later platforms were built in Atchafalaya, Barataria, Caillou, Terrebonne, and Timbalier Bays. Early platforms were virtually monopolized by Asians, who made their homes on the sites, supposedly to avoid attention from immigration authorities. According to legend, a great many individuals were smuggled into Louisiana by fishermen, who placed the aliens in barrels to bring them into the state unnoticed.

Shrimp-drying platforms were built of cypress planks on hand-driven pilings eight to ten feet high, which allowed air to circulate. After the shrimp were delivered to the platform, they were boiled in saltwater and spread on the wooden platform. There, they dried for one day if the weather was hot and sunny or several days under cloudy conditions. When rain threatened, they were covered with tarpaulins.

Courtesy Harry Cheramie

After drying, the heads and shells were removed by laborers who wrapped their shoes with cloths or sacks and "danced the shrimp," treading on them to remove the loose hulls. Small amounts could be beaten with a bundle of branches or a large homemade "flyswatter." The shrimp were then shaken on hardware cloth or poured from a height in a stiff wind to separate the loose shells from the meat.

At its peak, an estimated seventy-five drying platforms existed in Louisiana, of which probably the most well known was Manila Village in Barataria Bay. It was large enough to have its own post office. Established in 1884, Manila Village had nearly forty thousand square feet of wooden platforms for drying shrimp. In addition to the post office, the village had a general store, living quarters, and a storehouse, all set on wooden stilts over the bay's waters. Its population of nearly four hundred people included Filipinos, Chinese, Mexicans, and Anglos. Other shrimp-drying locations included Bassa Bassa, Bayou du Large, Cheniere Caminada, and Bayou Brouilleau.

Then, in 1922, two Louisianans, Fred Chauvin and Shelley Bergeron, received a patent for a mechanical dried-shrimp shelling machine. The

From William D. Chauvin Collection, Louisiana State Archives, Baton Rouge

device was essentially a revolving wooden drum with a beater and a screen sieve.

The Louisiana dried-shrimp industry peaked in 1929 with nearly eighty processors producing almost five million pounds. Two million pounds of that were shipped to China, and much of the remainder was exported to the Philippine Islands, Hawaii, and Latin America. The two largest exporting companies were Blum and Bergeron and Quong Sun. Shrimp destined for export were packed in wooden barrels, like many other commodities of the day, including salted and dried speckled trout, which were also produced on the drying platforms.

The shrimp-drying industry received another boost in the late 1950s with the invention of a gas-powered mechanical shrimp-drying machine. The device was patented in 1960 by Louis Blum, a family member of the firm of Blum and Bergeron. The mechanical shrimp dryer allowed shrimp to be dried indoors in smaller areas than those required for platform shrimp drying.

As late as 1962, twenty-three driers still operated, but most had

From the collection of Daniel Alario

abandoned the platforms. Time and weather had taken their toll. Many platforms were destroyed in the hurricanes of 1915 and 1926 and never rebuilt. Later hurricanes destroyed more, finally taking Manila Village in 1965.

In recent years, between 6 and 10 processors still produce dried shrimp. All of these use indoor dehydrators and rotating-drum shell-hullers. It takes almost 8 pounds of fresh shrimp to produce 1 pound of dried shrimp, so they have become a relatively pricey specialty food. Louisiana dried shrimp are now shipped to California, Hawaii, New York, and Canada, as well as to Asia.

Development of the Modern Shrimp Industry

From the 1870s, when commercial shrimping began in Louisiana, until the 1930s, shrimping was a part-time artisanal occupation—something to make it by between the real money to be made during winter fur trapping seasons. With no ice or refrigeration to keep the catch fresh, shrimping was limited to inshore bays and lakes. After being caught, shrimp were immediately offloaded at drying platforms located in the bays or brought by freight boats to shrimp canneries on higher ground. Shrimping was conducted with haul seines, and only later, between 1912 and 1915, were otter trawls introduced.

Changes began in the 1930s, and World War II cemented into place something similar to the modern shrimp fishery. Morgan City became Louisiana's commercial fishing boom town of the 1930s and '40s, partly due to its location at the southern end of the Atchafalaya River. However, the true driving force behind the establishment of Morgan City's role in the industry was probably due to the fact that Louisiana's offshore shrimp stocks were discovered from this port. In 1934, an

From the collection of Daniel Alario

out-of-state, forty-five-foot boat with a forty-horsepower engine put into port there carrying between forty and fifty barrels (a barrel of shrimp is 210 pounds) of very large shrimp caught in waters off St. Mary Parish. This discovery started an explosion of boat building and shrimping activity concentrated on the previously unshrimped offshore waters.

The Morgan City newspaper, the *Daily Review*, documented the growth of the fishery during the next decade and how World War II affected it. The first change the war brought was the requirement that by January 1942, all boatmen, including fishermen, carry identification cards with their pictures and fingerprints on them. By May, the war came home when three local trawlers brought in twenty-three survivors of a submarine attack on a merchant ship in the Gulf. In June, a group of Norwegian survivors of an attack was brought into port, followed by a lone lucky survivor from another torpedoed ship.

Also in 1942, the fishing industry began to feel the hardships of the war, as key men were being drafted for military service. In October, deferment of shrimp boat captains and engineers was sought. By February 1943, captains and engineers were allowed deferments from the draft ranging from ninety days to six months. Since it took one to

From Morgan City Archives

From Morgan City Archives

two years to train men for these jobs, the War Manpower Commission finally took action in March to exempt boat captains from the draft.

The exemptions were enacted, in part, because wartime shortages in the nation's meat supply made seafood supplies more important. Demand was so high that almost all seafood was shipped out of state, and in Morgan City itself, seafood became difficult for consumers to find. When seafood could be found, it brought a high price.

The high demand and price for shrimp caused the wartime Office of Price Administration (OPA) to step in and set maximum prices fishermen could be paid for shrimp. Previously, the prices that processors could charge for shrimp when they resold them were so low that processors couldn't afford to buy shrimp at the high prices fishermen were getting on the fresh market. Thus, the entire catch was being diverted to the fresh market. The fishermen's price ceiling was announced in September 1943:

Head-on (per 210-lb. barrel)		Headless (per lb.)	
Under 9 per lb.	$32.00	Under 12 per lb.	33¢
9-12 per lb.	$28.00	15-20 per lb.	28½¢
12-15 per lb.	$24.00	21-25 per lb.	24¾¢
15-18 per lb.	$20.00	26-30 per lb.	21½ ¢
18-25 per lb.	$17.00	31-42 per lb.	19¢
26-39 per lb.	$14.00	43-65 per lb	16½¢
40 and over	$11.00	66 and over	14¢

From the collection of Daniel Alario

From Morgan City Archives

From Morgan City Archives

From Morgan City Archives

Newspaper accounts record that shrimp 9-12 and 12-15 to the pound were the predominant sizes landed locally.

Other stresses on the shrimp industry appeared in 1943. Shortages of ice kept many boats at the dock. Parts were difficult to obtain and with few parts to service them, boats began breaking down. Often they were tied up for months waiting for repairs. The draft and high-paying factory jobs created a shortage of workers in seafood plants. The labor shortage became so serious that the Child Labor Bureau allowed fourteen and fifteen year olds to work at deheading and peeling shrimp.

In January 1944, Harvey Lewis, president of the Fishermen's Association, and John Santos, president of the Southwest Louisiana Shrimp Dealer's Association, teamed up to push the OPA to increase the price of shrimp 20 count and larger from $24 a barrel to $28 a barrel. A movement began in March to ensure all fishermen the same deferral from the draft as agricultural workers.

In spite of the problems, 1944 was shaping up to be a very good

From Morgan City Archives

From Morgan City Archives

shrimp year, with shrimp catches as high as forty barrels a trip in August. In November, a Morgan City shrimper, Ashley Galloway, brought in a catch of 147 barrels. November was called the "biggest month in history."

With the war nearing an end, Harvey Lewis traveled to Washington to receive approval for a plan in which the federal government would finance fishing boats for returning veterans. The war's end meant other changes as well. The first all-steel shrimp boat appeared in Morgan City in 1945. The ready availability of ice across the coast meant that trawlers could make longer trips. Effective refrigeration and freezing equipment resulted in more and more shrimp being frozen or shipped fresh and less shrimp being canned and dried.

Through all these changes, Louisiana's vast fishery for brown shrimp remained undeveloped. Though today the brown shrimp fishery is as large as the white shrimp fishery in pounds, as late as 1949, white shrimp made up 95 percent of the shrimp catch of the Gulf of Mexico

Courtesy Arthur Matherne

and south Atlantic states, with Louisiana producing two-thirds of that.

Though great advances would come in the next half-century, it was these years just prior to and following World War II that turned the Louisiana shrimping industry into more than just a part-time way to earn extra money for coastal residents. The progress during this period would enable the growth and supply the expanding demand of America's desire for this delectable seafood.

You Gotta Be Tough to Be a Commercial Fisherman

"You gotta be tough to be a commercial fisherman." That is fifty-two-year-old Byron "Rip" Despaux's answer to my question about how he could keep up the near round-the-clock pace on the last day of the three-day shrimping trip. He adds, "A lot of people don't know what we have to put up with. I've stayed awake as long as fifty-four hours straight without any sleep in order to make a decent trip."

Three days earlier, in his Barataria, Louisiana, home, Byron kisses good-bye his wife, Suzy, whom he describes as "my sweetheart since I was fourteen and she was thirteen and the only girl I ever loved." Minutes later he casts off the mooring lines to his immaculate thirty-five-foot Lafitte skiff and effortlessly pilots it into the ebbing tide on Bayou Barataria.

As he idles the boat south through the congested waterway, he explains that he comes from a family with generations of commercial fishermen before him. He bought his first boat when he was thirteen to fish bush lines for soft-shell crabs. Two years later he bought a larger Lafitte skiff. He began shrimp trawling full time in 1975, two years after finishing high school.

In 1980, at twenty-five years old, he and a partner invested several

hundred thousand dollars in the construction of an all-steel, ninety-five-foot vessel equipped with an on-board freezer, to shrimp offshore. In spite of the twenty- to thirty-day trips, life in the early years of the vessel was good. Expenses were reasonable, crews were easy for a profitable skipper to find, and shrimp prices were good.

Then, in 1983 the first great wave of imported farm-raised shrimp hit U.S. markets and shrimp prices began to slide downward, a trend that continues today. Concurrently, fuel prices began to rise, as did other costs. By 2005, when Byron sold the vessel, insurance costs alone were almost $22,000 a year. As profits declined, good crews became hard to find, a critical point for a vessel that used a three- to five-man crew, in addition to the captain. Of all of the challenges in owning an offshore shrimper, Byron says the worst was financial stress. Now, with a smaller boat, he has shed some of that stress and has rediscovered the joy of spending more time with his family.

Within a half-hour, we reach Bayou Rigolets, where our shrimping begins. As the slanting rays of the evening sun stream through a beautiful October sky, Byron uses electric winches to lower an aluminum skimmer net frame on each side of the boat into its near-horizontal-to-the-water fishing position. Suspended beneath each frame is the funnel-shaped skimmer net, a modern-day adaptation of the trawl concept. While setting up, Byron says that very few inshore commercial shrimpers still use the once-traditional trawls held open by otter doors, explaining that "skimmers are a more advanced way to fish."

"Skimmers produce better shrimp catches with less work and are more versatile than trawls," he adds. Byron has equipped each net with a "fish-eye" style bycatch reduction device (BRD, pronounced "bird") to release finfish from the net. "In most places," Byron says, "I can leave the frames out all night and continue to push [shrimpers with skimmers "push" their nets, while shrimpers with trawls "drag" theirs] while I empty the catch from the tails every hour or two. But here in Bayou Rigolets, the bottom is fairly dirty with bottom trash, so I'll probably have to pick the frames up every hour or so to be able to inspect the entire net for debris."

The debris that Byron refers to is primarily derelict crab traps and flotons, large, tough clumps of dead marsh grass stubble and roots that work their way from areas of dying marsh into deeper shrimping waters. Byron also explains that he will empty the skimmer tails more often when catching "more tender" young shrimp or when fishing for direct sale to the public, where appearance of the shrimp is critical.

By 6 P.M., the roaring diesel Caterpillar engine is driving the skiff and its skimmers southward at a steady 2 to 2½ knots. Byron explains that in clear-water conditions, shrimp can be very hard to catch during daylight hours and that skimming at night is usually much more productive. As he predicted, when the tails, referred to as "bags" by Byron, are emptied on deck at 7:30 P.M., shrimp numbers are sparse and the shrimp are small. Much of the catch consists of "jelly," the gelatinous transparent remnants of members of the jellyfish clan. By volume, shrimp outnumber finfish by about 5 to 1.

After separating out ten large male crabs to boil later and tossing back overboard eight large clumps of flotons, Byron shovels the catch into a salt box, a fiberglass box half-full of water with fifty pounds of salt dissolved in it. The high specific gravity of the saltwater allows dense shrimp to sink but forces lighter-bodied finfish and giant salvinia, a floating plant, to the surface, where Byron quickly dips them out with a dip net and flips them back into the water.

With most of the finfish removed, the remaining catch is dipped from the salt box and emptied onto a table-like picking box for final sorting. After the small crabs, clamshells, snails, and non-floating finfish are removed and tossed back into the water, Byron expertly sorts the shrimp into two piles, one of which counts 21-25 to the pound and the other, 50-60.

The shrimp are placed in baskets and then thoroughly washed with the heavy deck hose. The picking box, tools, and deck are then hosed spic and span. After draining, the shrimp are stored by size with a generous amount of ice in separate compartments built into the skiff. The total for the catch is five pounds of 21-25 count shrimp and about fifty pounds of small shrimp.

While Byron tended to his catch, the vessel steered itself on autopilot, steadily fishing. Every minute or two, while working, he would peer into the cabin at his radar and electronic plotter to be sure that he was

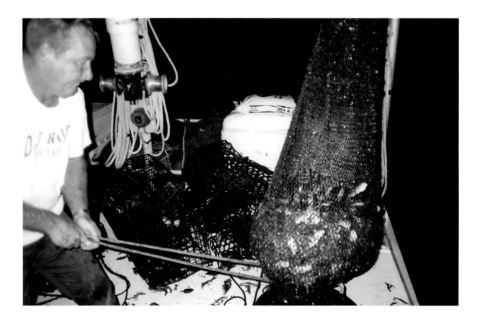

safely on course. With the catch now iced down, Byron returns to his station at the helm.

As he steers the vessel and peers into the night's blackness, he frets about catching small shrimp so early in the season. Shrimpers are paid more for larger shrimp than smaller shrimp. Also, like so many other shrimpers trying to survive, he retails many of his large shrimp directly to the public for a slightly higher price than he would receive at a wholesale buying dock.

"Very seldom do I get retail orders for small shrimp. Everyone wants big shrimp. But early in the [white shrimp] season, when shrimp are big, people are afraid to stock their freezers because they worry about losing them if a hurricane causes a power loss. Pretty soon, all we will be catching is 100 count shrimp, worth only fifty cents per pound, but everyone will want big shrimp."

By experience, shrimpers know that in the warm months of August and September, white shrimp grow rapidly and are in no hurry to leave the state's coastal lakes and bays for deeper offshore waters. As water temperatures become cooler, shrimp growth slows, and each cold front serves to successively flush white shrimp from inshore waters on strong outgoing tides, until the season closes in December.

True to Byron's earlier prediction, the next 1½-hour push yields better results, twenty pounds of large shrimp and about seventy pounds of small shrimp. As on the first push, he has to untangle two derelict crab traps from one skimmer and one from the other. The catch is cleaned and iced the same way as the shrimp from the first push.

And so it goes. All night, pushing south, then north, then south again, without stop, except to remove derelict crab traps from the skimmers and the endless twenty-to thirty-pound clods of floton from the catch. By the end of the eighth push, as dawn cracks at 6 A.M., I am punch-drunk from lack of sleep, but Byron seems as fresh as when we began.

The night's catch of an estimated 130 pounds of large shrimp and 640 pounds of small shrimp dissatisfies Byron, so he turns the vessel's bow southward to explore the potential of Little Lake. Only at 1 P.M., after three relatively unproductive daytime pushes, does Byron anchor the boat near a protected shore for a nap.

After a 3½-hour nap, he has the boat actively shrimping again. This time, the fishing day extends uninterrupted from 6:30 P.M., through the night, until noon the next day. Still his energy seems inexhaustible. Before turning in for his nap, Byron boils crabs for the evening's supper. Bleary-eyed, I abandon him for the comfort of the bunk.

Before 6 P.M., we are up and munching on crabs. By 6:45 P.M. we are shrimping again, following the same routine all night and well into the last morning.

Between pushes, Byron muses about his future and the future of the shrimp industry. "I guess that I will never get out of the fishing business. I shrimp all summer and crab with my son-in-law all winter. I love to hunt deer and ducks and fish for bass, but I have almost given that up. You got to make money when it is there to make.

"My grandfather, my father, and I all fished commercially, but it will end with me. My son, Byron, and I built this boat together, but I bought him out when he got tired of making a living in the summer

and starving in the winter. I encouraged him to get out of the business. I like the boat though. It is set up to be efficient. One man can operate it, although I usually have a deck hand with me."

He goes on, "Fishing is a way of life. I didn't want to be a cowboy when I was little; I always wanted to be a commercial fisherman. But, I don't think that it's got any future left in it. It is on a slow way out. Offshore shrimping is just about gone. Beach boats [medium-sized trawlers] are next. They are suffering from low shrimp prices, high fuel costs, and crewing problems. Inshore skiffs will be the last boats left. But even we are in trouble. A few years ago expenses before a three-day trip were $200 to $300 and shrimp prices at the wholesale dock were $2.50 to $3.00. Now expenses are $900 to $1,000 and we are getting $1.60 a pound for the same shrimp. When you only make a $200 profit for three days and nights of shrimping, it is disgusting. And to make things worse, shrimping is only seasonal.

"Very few in this next generation are getting into commercial shrimping," he glumly observes. "It costs over $100,000 for even a small skiff like this. A few young people are getting into crabbing, though. It costs less to get in the business and prices for the best of the catch are strongly up.

"In the future we will probably lose more boats in the shrimp fishery and the docks that buy shrimp for processing will probably be gone. Fishermen will have to focus more on catching larger shrimp and retailing them."

Byron's gloomy predictions are all the more concerning when one learns of his dedication to helping the fishing industry survive. He is chairman of the Board of Directors of the Louisiana Shrimp Association, a trustee for the Gulf and South Atlantic Fisheries Development Foundation, and maintains active membership in the Louisiana State Seafood Industry Advisory Board, the Jefferson Parish Marine Fisheries Advisory Board, the Louisiana Seafood Promotion and Marketing Board's Shrimp Task Force, and the Louisiana Farm Bureau Federation.

"Some of the problem is that fishermen are their own worst enemy," he observes. "They are never satisfied and always think someone is out to get them. The very worst thing is fishermen fighting other fishermen. That will always cause everyone to lose ground." He adds, "A big problem is that commercial fishermen want things to go back like they were twenty years ago, and that won't happen. They will have to learn to change, to go with the flow to survive."

Yet he characterizes commercial fishermen individually as the best of people, describing them as independent and able to make their own way. "They are trustworthy; almost always their word is their bond. And one fisherman can always count on getting help from other fishermen. They are like a big family."

After a long, silent moment staring over the endless water in front of him, he volunteers, "If someone offered me a job indoors, a less stressful job, I would take it." After another long moment, he contradicts his earlier statement, "I will never get out of the fishing business. We don't want to leave the bayou. Suzannah and I are family people and want to be around our family. Besides," he adds, with a twinkle in his eye, "I got a little thought in the back of my mind that it will get good again someday."

A Walk Through a
Shrimp Plant

A modern shrimp processing plant like DoRan Sea-Pak in Independence, Louisiana, is a maze of stainless steel constantly washed by sanitized water.

Before processing, shrimp are well iced and stored in plastic vats at 35°F.

Shrimp are washed three separate times and moved from one wash tank to another by conveyor.

After the final wash, the shrimp are routed into peeling machines. Within the peeling machines, water-bathed plastic rollers that work against each other catch the edges of the shrimps' shells and gently remove them.

The peeled shrimp are then moved through a separator to separate any loose shells, and an air blower beneath the conveyor belt blows away shell particles.

The shrimp are visually inspected by trained plant personnel. The bright blue conveyor belt aids visual inspection.

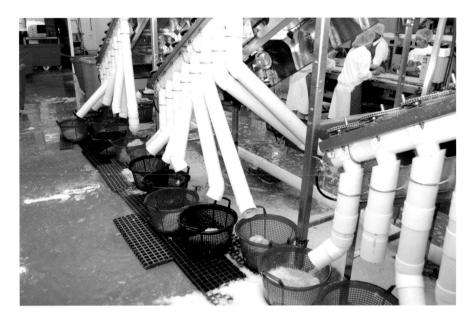

A size grader separates the peeled shrimp into six to eight different size categories, then the size-graded shrimp move by conveyor through one last inspection.

The shrimp are bagged in plastic bags and packed in five-pound waxed boxes.

The size count and other specifications of the shrimp are marked on each box.

The boxes are placed on racks before being blast frozen for eight to ten hours at -25°F.

Kinds of Louisiana Shrimp

Louisiana is blessed with a variety of shrimp. Not counting obscure deep-sea species, eight species may be found in commerce, although some are not common.

Common Name	Other Names	Scientific Name
White Shrimp	None	*Litopenaeus setiferus*
Brown Shrimp	Brazilian Shrimp	*Farfantepenaeus aztecus*
Pink Shrimp	Hoppers	*Farfantepenaeus duorarum*
Sea Bob	Six Barbes	*Xiphopenaeus kroyeri*
Roughneck Shrimp	Sugar shrimp, Blood Shrimp	*Trachypenaeus constrictis*
Royal Red Shrimp	None	*Pleoticus robustus*
Rock Shrimp	None	*Sicyonia brevirostris*
River Shrimp	None	*Macrobrachium ohione*

White shrimp, together with brown shrimp, comprise the overwhelming majority of shrimp caught and sold in Louisiana. Although both species are popular, white shrimp are almost always preferred by those people who express a preference. White shrimp are slightly more tender than other shrimp and their shells are a little softer and easier to peel. Though somewhat more expensive than other species, their ease of peeling after cooking makes them the shrimp of choice for recipes that cook shrimp in the shell, such as barbequed shrimp or boiled shrimp. Also, large white shrimp do not develop the slight iodine taste that other large shrimp do.

White shrimp can vary in color. Those caught at night or during certain seasons may be pinkish in color. Migrating white shrimp often will exhibit reddish-colored legs. These colors are no indication of quality and do not affect taste.

Differentiating white shrimp from other shrimp is nearly impossible once they have been deheaded. When the head is present on the shrimp, white shrimp may be identified by their very long maroon-colored whiskers, which are longer than their entire body.

Louisiana's inshore white shrimp season begins in August with catches comprising a mixture of sizes, including many larger ones. Smaller white shrimp are harvested in September, with sizes gradually

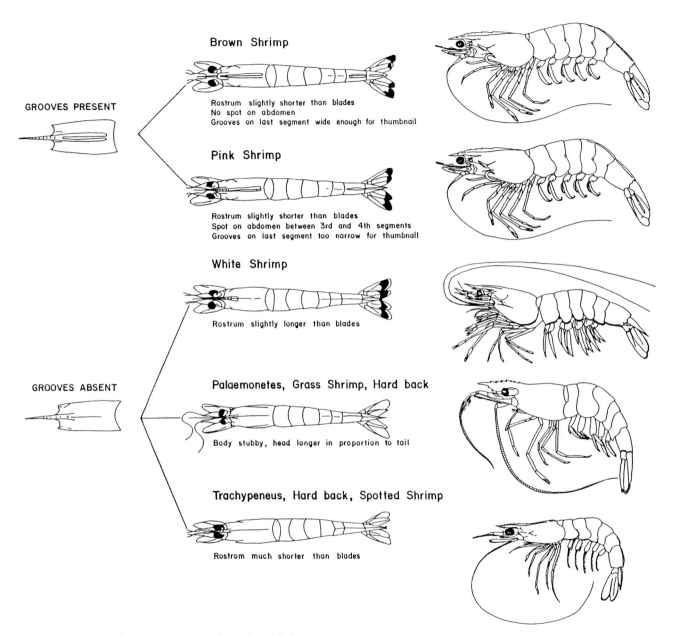

GROOVES PRESENT

Brown Shrimp

Rostrum slightly shorter than blades
No spot on abdomen
Grooves on last segment wide enough for thumbnail

Pink Shrimp

Rostrum slightly shorter than blades
Spot on abdomen between 3rd and 4th segments
Grooves on last segment too narrow for thumbnail

White Shrimp

Rostrum slightly longer than blades

GROOVES ABSENT

Palaemonetes, Grass Shrimp, Hard back

Body stubby, head longer in proportion to tail

Trachypeneus, Hard back, Spotted Shrimp

Rostrom much shorter than blades

Courtesy Texas Parks and Wildlife Department

increasing into October, when cooler weather slows their growth rate and smaller shrimp are more often harvested. Larger white shrimp are available in smaller amounts on a more or less continuous basis from offshore waters.

Brown shrimp, in most years, produce the largest harvest by volume of any species. Although they will individually grow as large as the largest white shrimp, most brown shrimp are harvested in smaller size classes.

Brown shrimp are firmer fleshed than white shrimp and when they reach a large size, they sometimes develop a slight iodine taste. As brown shrimp grow, they move farther offshore into more saline waters. There, they feed on a very common bottom-living creature, the acorn worm, which has a noticeable iodine-like odor that it imparts to shrimp when eaten. Commercial shrimpers often refer to these shrimp as "iodoform shrimp." This taste may or may not be noticeable to the average palate.

Brown shrimp colors will vary depending on size and where they are harvested. Small shrimp in inshore waters are often gray, although in some low-salinity waters they may be nearly black in color. Interestingly, while commercial shrimpers usually prefer to eat white shrimp rather than brown shrimp, most praise these low-salinity blackish-colored brown shrimp as having an exceptionally sweet taste.

Brown shrimp harvested in nearshore Gulf of Mexico waters are usually tan in color. Large brown shrimp harvested in deep offshore waters may be a rich brown to almost red. Generally, the browner the color, the more pronounced is the iodine taste. Cooked brown shrimp flesh will generally be very slightly pinker than cooked white shrimp.

Brown shrimp are quite small when the inshore season opens in Louisiana each May. Larger brown shrimp become available later in the season, before it closes in July. Large brown shrimp are harvested year round in offshore waters.

Louisiana's harvest of pink shrimp is small compared to its white and brown shrimp harvests. Most harvesting occurs before or at the beginning of the spring shrimp season in May and consists of larger shrimp. Because of their size and the fact that most are caught in or near the Lenten season, when local demand is high and fresh shrimp supplies are low, most pink shrimp are sold through local fresh retail markets. Occasionally, in years of good production, some may be processed. Pink shrimp landings may increase or decrease by a factor of thirty from one year to the next.

Though they are not harvested in large amounts in Louisiana, pink shrimp appear quite often in retail seafood specialty stores, especially during the mid-winter months, when the pink shrimp harvests peak in southern Florida and Louisiana's supplies of fresh shrimp are scarce.

Pink shrimp resemble brown shrimp but will have a dark, pigmented spot on each side halfway down the tail. They are not dependably pink in color. Quality-wise, pink shrimp more closely resemble brown shrimp than white shrimp. They are somewhat firmer than white shrimp and also taste more like brown shrimp than white shrimp.

Sea bobs are a marine shrimp that get very little respect in the marketplace, in spite of being landed in respectable quantities. This may be due to their small size, as they seldom grow much larger than 70 shrimp to the pound.

Sea bobs have an excellent taste and are easy to peel, making them good for shrimp boils, in spite of their size. Their storage life is somewhat shorter than other shrimp, and size for size, they are usually less expensive than other shrimp.

Most sea bob catches are made in December, January, and February, with smaller and sporadic catches in August and November. Shrimp processors make extensive use of sea bobs for peeling and freezing and for drying. Sea bobs never appear fresh in retail markets. Knowledgeable consumers purchase them directly from fishermen or shrimp-processing plants.

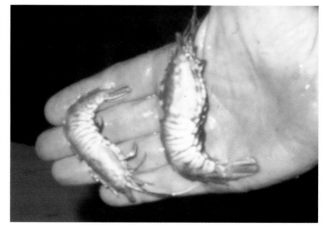

Roughneck shrimp are small, typically ranging from 50 to 70 to the pound. They have an attractive reddish pink coloration on their legs and head (hence the alternative name "blood shrimp")

and a larger-than-expected, dark-colored vein in their tail. The meat turns white when cooked, but because of its vein color, it lends itself better to battering and frying than to boiling, broiling, or use in sauces.

Roughneck shrimp freeze well but have a limited fresh shelf life. They have a large head and thick shell and peel easily. The flesh is tender but tastes similar to brown shrimp. Like sea bobs, these shrimp are slightly less expensive, count size for count size, than white, brown, or pink shrimp. They are often peeled and frozen by shrimp-processing plants and can be challenging to find fresh. Production of these shrimp is highly unpredictable from year to year. Most catches are made in March, April, and May.

Royal red shrimp are a medium-sized shrimp found in very deep waters off Louisiana's coast year round, but they are produced only on an occasional basis. Market demand for royal red shrimp is weak and production costs are high due to the deep waters they inhabit.

Royal red shrimp have a large head. Their flesh is more tender than other shrimp produced in Louisiana and shrinkage after cooking is higher than for other shrimp. These shrimp are bright red in color, are very attractive, and have an excellent taste.

Rock shrimp occur in moderate numbers off Louisiana's coast, but they are not produced regularly, in spite of a growing national market demand. While fairly good incidental catches have only been made from July into September, rock shrimp may also be abundant in other months.

Rock shrimp have thicker, harder shells than other Louisiana shrimp. In fact, the shell more closely resembles that of a crawfish, except that it is rougher and more pitted. They are a medium-sized shrimp, ranging between 30 and 60 shrimp to the pound. The taste of rock shrimp is very good and much richer than that of other marine shrimp.

Rock shrimp are often caught on sandy bottoms, so they may hold a good bit of sand in the crevices of their shell and in the vein through the tail. They should not be frozen for an extended period without the vein having been removed, as the vein will, with time, cause discoloration and an objectionable taste and odor. The sand on the exterior of the shell can be washed off with a good rinsing if the shrimp are to be cooked with the shell on.

River shrimp are a freshwater shrimp, very common in the Mississippi and Atchafalaya Rivers. They have a soft-textured flesh and a larger head than saltwater shrimp, much like the cultured Malaysian giant prawn, to which they are closely related.

The Forgotten Shrimp

River shrimp are part of Louisiana's forgotten seafood heritage. These tiny (70 to the pound is large), but succulent morsels usually come from the Mississippi River, although they are also common in the Atchafalaya River. River shrimp are not only small, but they also have a larger head relative to their tail size than saltwater shrimp. The flesh is more soft-textured than that of marine shrimp and is often described as "sweet tasting" by aficionados.

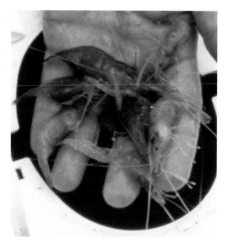

While few people eat them or even know of them today, the French adventurer Antoine Simon Le Page du Pratz wrote of them in 1758: "The Shrimps are diminutive crayfish; they are usually about three inches long, and of the size of a little finger. Although in other countries they are generally found in the sea only, yet in Louisiana you will meet with great numbers of them more than an hundred leagues up the river."

The Cajuns of the German Coast and farther inland along the Mississippi River (St. Charles Parish north to Pointe Coupee) made heavy use of them in their cookery. River shrimp also had a loyal following in Baton Rouge and New Orleans. The well-known Louisiana biologist Percy Viosca said of them, "This is a species which has always been considered a great delicacy and a food of gourmets in New Orleans."

In the July 1915 issue of *Louisiana Grocer* magazine, discussion was made of river shrimp.

Shrimps, too, are one of the greatest delicacies of the New Orleans markets; lake shrimps, large and fine; and better than these, river shrimps, small but toothsome to the last degree. All along the river front are men who call themselves "swimpers," and who have forty or fifty traps, each, which they sink into the water at the end of long ropes, and visit every hour. It is the product of this industry which is brought to the table in a crystal bowl, pink and resigned, and half covered with cracked ice; and the chef who has prepared them knew just how much salt and red pepper to use, so that they come forth perfectly seasoned.

The traps referred to were constructed of wood, usually cypress, and were called shrimp boxes.

The commercial river shrimp fishery saw its most productive years between 1933 and 1939, with a peak of 2½ million pounds landed in 1936. The price range was seven to nine cents per pound. For comparative purposes, commercial landings of saltwater shrimp in 1936 were 53 million pounds. Modern landings of saltwater shrimp are about 100 million pounds.

After 1941, commercial river shrimp landings began a rapid decline, but prices received for them increased markedly, forced up to surprising

Courtesy Malcolm Comeaux

levels by strong demand for what was being caught. Wholesale prices peaked in 1953 at 50 cents per pound before beginning to decline. As supplies then began to decline, prices began to move up again. By 1975, river shrimp were sold at retail markets in Baton Rouge for $1.99 per pound, placed in a bin right next to much larger saltwater shrimp selling at $1.29 per pound. The river shrimp rapidly sold out. The changes in the fishery were attributed to the Great Depression in the 1930s, followed by labor shortages after World War II began. Whatever the case, the fishery has never returned to 1930s volumes.

The last recorded commercial landings of river shrimp in Louisiana were made in 2000, with 155 pounds. None have been recorded since, although it is very likely that some catches are made and sold under the radar directly to consumers. Recreational fishermen still fish for river shrimp for personal consumption and for use as fish bait.

River shrimp are not sold in retail markets, even on the German Coast of the Mississippi River. One's best chance of finding river shrimp is to nose around river towns in St. James, Ascension, and Iberville Parishes and ask a lot of questions. Then, when you find a source, protect it.

Shrimp Life Cycle

White, brown, and pink shrimp are estuarine-dependent tropical shrimp. As adults, they spawn in waters extending from the beaches to several miles offshore. Each female will produce up to one million eggs, which hatch in twenty-four hours. The larvae go through eleven stages, shedding their shells for each stage.

As larvae, they feebly swim and float around as plankton from the top to the bottom of the sea. After the eleven larval stages, during which they look nothing like shrimp, they molt into miniature versions of adults, called post-larvae. At the second post-larval stage, when they are about as long as the width of a human fingernail, they hitch a ride on tides from the sea into estuaries.

Once in the marsh, they settle to the bottom and begin feeding and growing as juvenile shrimp. In a couple of months they grow to be 4- to 4½-inch subadults and move from the marshes into larger lakes, bays, and sounds.

After several weeks or months in this habitat, they move out through the passes to the sea on strong outgoing tides. There, they mature and spawn. Less than 2 percent of all eggs will survive to adulthood and few shrimp live beyond one year.

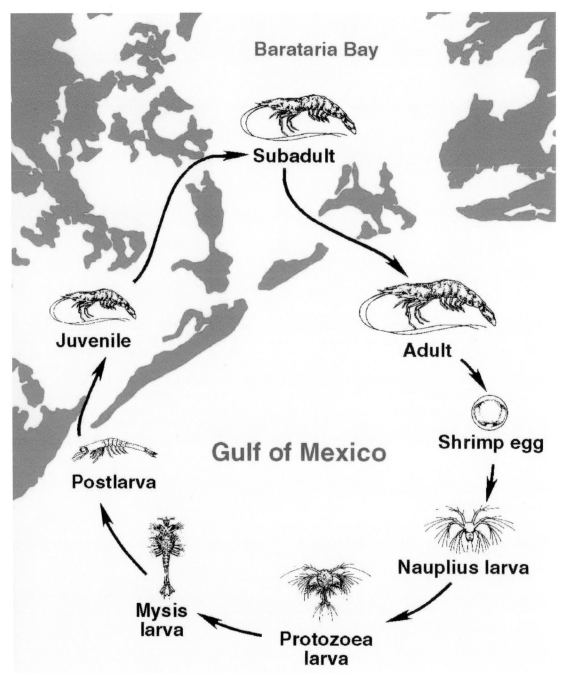

Barataria Bay

Subadult

Juvenile

Adult

Gulf of Mexico

Shrimp egg

Postlarva

Nauplius larva

Mysis larva

Protozoea larva

Courtesy Louisiana Sea Grant College Program

Soft-Shell Shrimp?

Soft-shell crabs are a well-known, but expensive seafood staple in Louisiana. A little searching can locate soft-shell crawfish as well, but soft-shell shrimp? Yes, they do exist, but to get them one must combine detective skills with the power of persuasion.

All crustaceans—crabs, shrimp, crawfish, and lobsters—are locked in a hard shell called an exoskeleton. This shell cannot expand, even for thin-shelled crustaceans such as shrimp. As the animal grows, it fills its shell but it can't grow larger until it sheds the old shell in a complex process called ecdysis, or molting. The new, very soft shell takes several hours' immersion in water to harden.

Fisher folk have some control over this process in crabs and crawfish, which show color and behavioral changes prior to molting. These pre-molt animals are placed in indoor chambers with good water flow to allow them to shed their old shells in a controlled environment.

Immediately after they shed, the crustaceans pump up their new soft shells with water to expand them by as much as 30 percent. When the expanded shells harden, the animals have new, larger shells in which to grow. During the short period between the expansion of the shell and its

hardening, the animals must be harvested to be consumed as soft-shells.

While crabs and crawfish are resilient enough to be held alive before and during the shedding process, shrimp are far too fragile for this treatment, as they will die during any period of time out of the water for transportation to a shedding facility. Soft-shell shrimp can only be obtained by persuading a shrimp fisherman to sort the soft-shells from the rest of the catch and save them for you rather than for themselves. Shrimp seem to shed in numbers only at night, mostly the blackest or most overcast of nights, and they are most often absent from the large catches of migrating shrimp made in passes at night.

Soft-shell shrimp will be quite expensive, but they make a most impressive dinner presentation. Few people besides commercial shrimpers have tried them. Preparation and cooking is easy. Simply snip the snout off behind the eyes with a pair of scissors, then batter and fry them.

Shrimp Product Forms

Shrimp can be purchased in a huge array of product forms: fresh head-on; fresh headless; frozen head-on; frozen headless; frozen peeled and deveined; frozen cooked, peeled, and deveined; frozen cooked, peeled, and undeveined; frozen breaded; and dried.

Fresh head-on shrimp are completely unprocessed, sold just as mother nature produced them, and hopefully well-iced. Frozen headless shrimp are usually sold in five-pound boxes or poly bags

From William D. Chauvin Collection, Louisiana State Archives, Baton Rouge

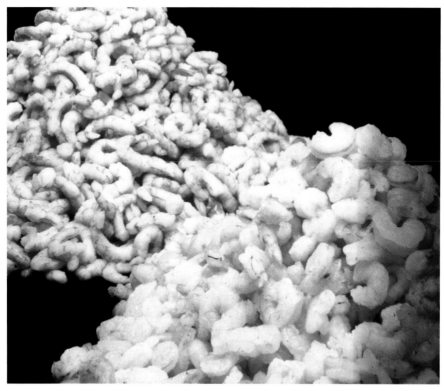

From William D. Chauvin Collection, Louisiana State Archives, Baton Rouge

and are very often referred to as "green headless" shrimp. "Green" does not refer to any particular species, but rather is a traditional term meaning "unprocessed."

Peeled and deveined shrimp are often abbreviated as "p & d" and peeled and undeveined shrimp as "p.u.d." Both are available cooked and uncooked and are most often sold in poly bags. Each step of processing causes weight loss and a corresponding price increase. Shrimp sold as p.u.d. are more expensive than green headless. In turn, p & d shrimp are more expensive than p.u.d. shrimp. Cooked shrimp are more expensive than raw shrimp because shrimp lose fluids during any cooking process, including boiling.

Dried shrimp are a high-quality specialty product unique to Louisiana. While some are sold in Louisiana, primarily as a snack food, the quality of Louisiana dried shrimp means many of them are exported to Asian markets. It takes an average of 7.69 pounds of fresh head-on shrimp to produce 1 pound of dried shrimp. This tremendous weight loss means that dried shrimp are quite expensive.

Canned shrimp are no longer produced in the United States. From two dozen shrimp-canning "factories" in the mid-1970s in Louisiana and Mississippi, the business was down to one plant by 2005. This plant was destroyed by Hurricane Katrina and is not likely to be replaced. The extinction of this once most important type of processing is due to cheaper Asian imports and a decline in canned food consumption by Americans. All canned shrimp available on store shelves today are imports. Typically, they have a softer, more crumbly texture than did domestically canned shrimp.

Shrimp Sizes and Serving Sizes

Since the price of shrimp increases as shrimp become larger and a bewildering array of sizes is available, a consumer is often left wondering if they have made a good buy. While the use of the terms such as large, medium, and small are fine for home-style recipes, the only way to be certain of value in purchases is by buying shrimp by count size. Count size simply refers to how many shrimp are in a pound. The size categories for head-on, shell-on shrimp, starting with the largest are as follows: under 10 (U/10), 10-15, 16-20, 21-25, 26-30, 31-35, 36-40, 40-50, 50-60, 60-70, 70-80, 80-100, 100-120, and 120+.

Shell-on headless shrimp are sized similarly, but they are not sized smaller than 80-100 count to the pound. The following table shows the nomenclature system of sizing headless shrimp and the corresponding

From William D. Chauvin Collection, Louisiana State Archives, Baton Rouge

count for each size. Headless shrimp are typically sold frozen in five-pound waxed boxes and referred to as "green headless" shrimp. The term "green" does not refer to color or species, but rather to the fact that the shrimp have not yet been further processed.

Shrimp Size	Count per lb.	Shrimp per 5-lb. box
Extra Colossal	U/10	40-49
Colossal	U/12	50-59
Colossal	U/15	60-74
Extra Jumbo	16-20	75-97
Jumbo	21-25	98-120
Extra Large	26-30	121-145
Large	31-35	146-173
Medium Large	36-40	174-190
Medium	41-50	191-240
Small	51-60	241-290
Extra Small	61-70	291-340

The nomenclature system does not appear in law, so a great deal of variation exists, always with the opportunity for creative pricing, hence the value of buying by count size. One of the more humorous examples of creative pricing we have seen are signs on shrimp peddlers' trucks advertising "extra-medium" shrimp. Such language apparently sounds more attractive than offering only medium shrimp, but if the shrimp are larger than medium, shouldn't they be marketed as medium-large?

Though the terms large, medium, and small serve well for non-commercial cooking, when knowing the precise number of shrimp to a package is not necessary, restaurateurs need to be more exact about the count of their shrimp. They plan menus and calculate product costs based on how many individual shrimp are in a box or package.

Besides being available in size counts, frozen headless shrimp may be purchased as pieces. Pieces are less than perfect tails, which may have been pinched by a crab during harvest or otherwise damaged. No rigid standards exist for grading pieces, but some processors use large (0-50 pieces to the pound), medium (50-75 pieces), and small (over 75 pieces). Because pieces are priced lower and perceived as a bargain, they are often in short supply.

Peeled shrimp are graded in finished size counts to the pound, of 31-35, 36-40, 40-50, 50-60, 60-70, 70-90, 90-110, 110-130, 130-150, 150-200, 200-300, and 150+. In other words, peeled shrimp are graded after they have been peeled. Most, but not all, larger peeled shrimp in the sizes 10-15, 16-20, 21-25, and 26-30 are not graded in finished (end product) counts, but rather using "from" counts. This means that the end peeled product was derived from a certain size of shell-on shrimp tail before peeling.

The average number of shrimp per serving size for the most popular size counts of peeled and shell-on headless shrimp are shown below.

Peeled Serving Size Information

Count Per Pound	21-25	26-30	31-35	36-40	41-50	51-60	61-70	71-90	91-110	110-130	130-150	150-200	150+
Average No. Shrimp Per Serving	6	7	8	10	11	14	16	20	25	30	35	44	56

Shell-On Headless Serving Size Information

Count Per Pound	U/12	10-15	16-20	21-25	26-30	31-35	36-42	43-50	51-60	61-70	70-80	80-90	90-100
Average No. Shrimp Per Serving	3	4	6	7	9	11	13	15	18	21	24	27	30

From William D. Chauvin Collection, Louisiana State Archives, Baton Rouge

From William D. Chauvin Collection, Louisiana State Archives, Baton Rouge

Shrimp Weight Conversion Chart

It seems that when a recipe calls for headless shrimp, all you can find in your market is head-on shrimp or vice versa. You need to somehow figure how much shrimp to buy to get the amount called for in the recipe. Use the table below to get your answer.

Product	Percent remaining from heads-on weight	To convert to heads-on weight multiply by
Brown Shrimp (headless)	62.1	1.61
White Shrimp (headless)	64.9	1.54
Pink Shrimp (headless)	62.5	1.60
Sea Bob (headless)	65.4	1.53
Royal Red Shrimp (headless)	55.6	1.80
Peeled, raw (average)	49.0	2.04
Peeled, cooked (average)	31.9	3.13
Dried (average)	13.0	7.69

Where to Buy Shrimp

Shrimp are widely available, but price and quality also vary widely. Perhaps the most convenient place to buy shrimp is a grocery store. But grocery stores have three possible drawbacks: they can be expensive, often their shrimp are not local in origin, and you can't be sure that the shrimp have not been treated with sodium bisulfite or phosphates during handling and processing.

Another option is a retail seafood specialty store that handles only seafood. These vary in quality and service. Some purchase their shrimp directly from fishermen who fish exclusively for the retail

From William D. Chauvin Collection, Louisiana State Archives, Baton Rouge

From William D. Chauvin Collection, Louisiana State Archives, Baton Rouge

trade; others purchase their shrimp from dockside shrimp buyers and shrimp processors. Many do both. The best strategy is to develop a personal relationship with a store operator whom you trust and stick with him or her.

Consumers who want to stock their freezers should probably consider purchasing their shrimp from dockside shrimp buyers, shrimp processors, or directly from commercial fishermen. Dockside shrimp buyers and most shrimp processors are located in coastal commercial fishing communities. Virtually all of them will sell shrimp to the public. Do not expect bargain-basement prices, but rather, reasonable retail prices. Other than price, perhaps the major advantages in purchasing shrimp at this level are freshness and the availability of a variety of sizes, product forms, and species.

Consumers with the most dedication to quality may opt to go right to the source, the commercial shrimper. Do not expect to buy shrimp from fishermen for the same low price that they sell to the wholesale dock or processor. Although some fishermen are more quality conscious than others, most who sell to the public reserve their freshest and best catches for such sales. To get the freshest shrimp possible, shrimp should be purchased from shrimpers who make trips of three days or less in duration. Large shrimp vessels often make longer trips and unless they

are equipped with an on-board brine freezer, the shrimp catch may be treated with sodium bisulfite to prevent undesirable color changes.

Locating a fisherman demands a little effort, unless you have a referral from a friend. This usually involves a trip "down the bayou" to a fishing community. A few questions to almost anyone in these small communities will usually yield several names. It is essential to place an order with your commercial fisherman before his trip. Trying at random to catch a shrimper at the end of a trip is difficult.

Just as good as buying direct from a fisherman at the end of his trip is buying shrimp at an organized farmers' market that allows the sale of seafood. Such markets have strict rules that limit venders to selling only their own catch. This is done to preclude professional peddlers from selling shrimp of unknown origin in the markets. Farmers' markets are ideal for purchases of ten pounds or less. Larger quantities to stock your freezer are best special-ordered from a fisherman.

Two additional sources exist, both with drawbacks. You can catch your own shrimp. But, unless you live in an area where you can cast net for them, sport trawling is hard work and almost always a money-losing effort.

Buying from shrimp peddlers who set up a truck on the side of the road may be disappointing for several reasons. Do not be fooled by the white boots they often wear; they are usually not fishermen, but rather professional peddlers. The only exception to this is when a fisherman's family member, usually a wife, makes an effort to improve their family income with direct sales. Commercial fishermen almost never have time to peddle their catch themselves.

In general, peddlers' prices are not the best deal possible, as they purchase their shrimp from seafood dealers, often several steps removed from the fisherman. Unquestionably, the most abused shrimp we have seen offered for retail sale in thirty years of observation have been from the ice chests of peddlers.

Fresh Seafood from Farmers' Markets

Fruit and vegetable lovers have long known that organized farmers' markets are among the best sources of fresh and unusual local produce, direct from the producer to the consumer. Until recent years, buying seafood at Louisiana farmers' markets was not an option. Now farmers' markets from Baton Rouge south offer fresh seafood, sold by the commercial fishermen themselves, on a regular basis. Louisiana markets with seafood vendors are listed below.

Ascension Fresh Market
Sorrento, Ascension Parish
www.ascensionfreshmarket.org
(225) 675-1752

Crescent City Farmers' Markets
2 locations, New Orleans, Orleans Parish
www.crescentcityfarmersmarket.org
(504) 861-5898

German Coast Farmers' Markets
Luling and Destrehan, St. Charles Parish
http://germancoastfarmersmarket.org
(e-mail) wandrus@bellsouth.net

Gretna Farmers Market
Gretna, Jefferson Parish
(504) 362-8661

Mid-City Green Market
New Orleans, Orleans Parish
http://midcitygreenmarket.org
(504) 344-4429

Red Stick Farmers Markets
3 locations, East Baton Rouge Parish
http://breada.org
(225) 267-5060

Upper Ninth Ward Farmers' Market
Upper Ninth Ward, Orleans Parish
(504) 482-5722

Vietnamese Farmers' Market
New Orleans East, Orleans Parish
(504) 861-5898

Courtesy Kim Chauvin, Mariah Jade Shrimp Company

Country of Origin Labeling

Where shrimp are caught is important to many consumers. Banned antibiotics and additives have been found in some seafood imports. Some people prefer to "think global and eat local," believing that long-distance transport of food increases its carbon footprint on the earth. Others are patriotic and prefer "made in America." Many feel that local products—fruits, vegetables, and seafood—just taste fresher and better.

Whatever one's reasons for buying local, the passage of country of origin labeling (COOL) requirements for fresh and frozen seafood by Congress in 2002 has made determining where seafood comes from easier. Under the law, retailers—except for seafood specialty stores and restaurants, which are both exempt—must inform consumers by a clear label or sign where the product is from and whether it is wild caught or farm raised. To be labeled as a product of the U.S., farm-raised seafood must be hatched, raised, and processed in the U.S. Wild-caught seafood must be harvested from U.S. waters or by a U.S.-flagged vessel and processed in the U.S.

Unfortunately, quite a few exemptions from COOL do exist. Cooked, cured, smoked, canned, or surimi-type products are exempt from labeling. When two or more different seafoods are mixed together, they don't need to be labeled, even if they are all imports. Also exempt from labeling requirements are substantially modified seafoods such as breaded products, marinated products, soups, sauces, sea-food salads, and cocktails, sushi, and pâté.

Recognizing High-Quality Shrimp

Seafood seems to carry a certain mystique, intimidating many would-be buyers. However, recognizing high-quality shrimp is really quite simple. Just look for the following indicators:

- The nose knows! Don't be bashful. Ask the vender for a shrimp to smell. Fresh shrimp have no odor, or better yet, they smell like the ocean or seaweed. A fishy smell means that the shrimp are on the short road to decay. The stronger the smell, the closer they are to putrefaction. Never buy shrimp with a chlorine smell. They may have been rinsed in a chlorine solution to disguise other problems.

- For head-on shrimp, the head should be firmly attached to the bodies. One of the first signs of trouble in old shrimp is loose heads.

- Fresh shrimp, even pink shrimp, should not have opaque pink or pink-orange-tinted bodies. Such shrimp are actually beginning to cook themselves. Some shrimp species, such as royal red or large brown shrimp, will have naturally reddish bodies but will be translucent when fresh, not opaque. Note also that some very fresh shrimp will have pink swimming legs (swimmerets) on the bottom of the tail. Pink swimmerets are not a sign of deterioration.

- Melanosis, a black discoloration at the joints or segments of the tail, is not itself a sign of decay, but rather a sign of mishandling at harvest. The color is due to an enzymatic reaction at harvest. Such shrimp were left in direct sunlight too long or were not properly washed after sorting. Sunlight triggers the melanosis reaction. It may be best to avoid such shrimp, as they are unsightly, at best.

- The shells of shell-on shrimp, except rock shrimp, should be smooth and not pitted or gritty. Pitted shells are a sign

Courtesy Russ Miget, Texas A & M Sea Grant

of improper overexposure to sodium bisulfite, a chemical used to prevent melanosis. Sulfites are an issue for sulfite-sensitive people.

- Peeled raw shrimp should not be overly translucent or have a glassy or slimy feel when thawed. These are signs of overexposure to sodium tripolyphosphate, an additive that can cause shrimp to absorb water.

Be aware that some retail specialty stores and most mobile truck venders sell shrimp from bins or ice chests of ice water. This is a deliberate attempt to add weight to the shrimp, which they will sell by the pound. Substantial amounts of water may be retained between the shell and the body long enough to be weighed. This water is also a quality concern. As ice melts and drains, it washes bacteria down with it. If it has nowhere to go, the water simply serves as a cold bacterial bath for the shrimp.

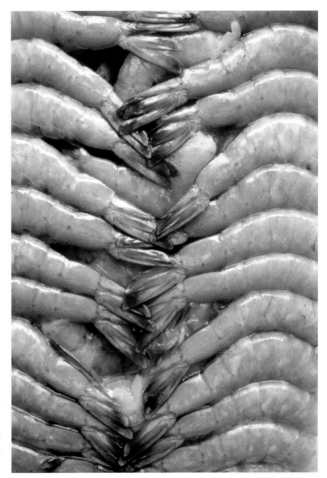

From William D. Chauvin Collection, Louisiana State Archives, Baton Rouge

Bad Bugs and Shrimp

Shrimp are one of the safest of all animal proteins to consume. Unlike finfish and some meats, shrimp carry few parasites that can be passed on to humans. The same is true for toxins from the marine environment. Paralytic shellfish poison, ciguatera, scombrid toxin, and pufferfish toxin are never found in shrimp.

Shrimp, like any other food, can cause illness if mishandled, however. The most serious illness-causing problem occurring during the home cooking of seafood involves cross-contamination. At least 12 species of *Vibrio* bacteria have been implicated in human illness and these bacteria are among the most common organisms in surface

From William D. Chauvin Collection, Louisiana State Archives, Baton Rouge

waters of the world. Two, *Vibrio parahaemolyticus* and *Vibrio cholerae* cause seafood-related illnesses, almost always by cross-contamination. Most often, this occurs when boiled seafood is temporarily placed back in the unsanitized containers that they were held in raw.

Of the two, *V. parahaemolyticus* is by far the most common and almost always the culprit when a number of people become ill four to ninety-six hours after a seafood boil. Symptoms include diarrhea, cramps, nausea, vomiting, chills, and fever and usually pass within 2½ days. Hospitalization is seldom needed. The bacterium does its damage by attaching itself to the victim's small intestine and excreting a toxin. More *V. parahaemolyticus* infections occur in the warmer months of the year than in cooler months. This bacterium is free-living and very common in all U.S. estuarine and marine waters, so any container that held raw seafood probably has the organism on it. Some containers, such as plastic ice chests, can be sanitized and reused; others, such as wooden crates, cannot.

V. cholerae is less common but poses a more serious threat to those infected. This bacterium causes the cholera that was so common and feared in the U.S. in the 1800s. Contaminated drinking water was the source of most of those illnesses. By 1911, good sanitation had made cholera outbreaks a thing of the past. Then, beginning in 1973, multiple cholera illnesses were noted in southwestern Louisiana, all of them traceable to cross-contamination in home-boiled crabs. After several years, the number of human illnesses abated, but experts still find the organism in the U.S. coastal waters.

Cholera is serious and may cause death. Symptoms are similar to *V. parahaemolyticus* infections but much more violent. The patient may lose gallons of fluids within a day or two. Shock, followed by death, can occur hours after symptoms begin, although eighteen hours to several days is the more usual course. Massive re-hydration is necessary, as 50-60 percent of untreated cases will result in death.

Food Additives in Shrimp

Compared to many foods, few additives are used in shrimp harvest and processing. Sodium bisulfite is a chemical that may be used as a mild dip by fishermen at sea to prevent melanosis, or black spot. Such treatment is legal and harmless to most people, except, in theory, those who are sulfite sensitive. In spite of many years of sulfite use on shrimp without a single documented case of a reaction to sulfites, the possibility must be noted. Many shrimp, especially those produced by short-trip fishermen who shrimp for direct-to-public sales, are never treated with sulfites.

Most controversial is the use of sodium tripolyphosphate (STP), especially for the processing of peeled shrimp. Proper phosphate use is legal and the U.S. Food and Drug Administration has included phosphates on its "generally recognized as safe" list. Used in minimal amounts they can protect the texture and flavor of mechanically peeled shrimp by preventing moisture loss.

However, since phosphates bind water to seafood, the temptation to add more than the maximum allowable level of phosphates in an attempt to add water weight to the product exists. Overuse of phosphates is not a human health issue but rather a problem of weight fraud and table quality. Overtreated shrimp may feel slippery and the surface of the product will become transparent. Overtreatment can gel the flesh, cause it to look unpleasant and lose texture, and make the shrimp nearly impossible to cook to an opaque appearance. Such shrimp also have a rubbery rather than firm quality after cooking. It is best to not buy from sources where one has previously found STP problems.

Black Spot on Shrimp

Occasionally, in shell-on shrimp, both headless and head-on, a very dark, almost black discoloration can be seen in the joints between the segments of the tail shell. In severe cases the tail may be more black than light colored. This is usually called "black spot," or more properly melanosis.

While it looks terrible, and it is considered a defect in the marketplace, black spot is not caused by high bacteria levels and spoilage. In fact, large numbers of bacteria actually slow down that development because they use up the oxygen needed for black spot formation. Black spot formation is a chemical reaction similar to sun tanning that involves the shrimp's own amino acids, so the longer that shrimp are exposed to sunlight on the deck of the shrimp vessel, the faster they will get black spot in the vessel's hold.

Courtesy Michael Moody, Louisiana State University

Courtesy Russ Miget, Texas A & M Sea Grant

Shrimpers work to prevent black spot by quickly handling the shrimp when they are exposed to sun, thoroughly washing them, and then storing the product in a good quality melting ice. The melting of ice is important because runoff from the ice washes away the black spot agents. Some shrimpers also use chemicals, especially sodium bisulfite, often called "shrimp powder" or "dip." This material works by binding the oxygen needed for the black spot reaction.

Unfortunately, a small number of people are sulfite sensitive and the presence of the chemical in some foods can cause allergic reactions. Although it is required that the presence of any additive be noted on labels, it is possible that shrimp treated on a vessel may come in contact with untreated shrimp during processing. Sodium bisulfite is most often used on vessels that make trips longer than three or four days. Sulfite-sensitive consumers can avoid sulfites by opting to buy their shrimp directly from short-trip (trips of three days or less) commercial fishermen.

A product gaining attention for the prevention of melanosis is 4-hexylresorcinol, marketed under the trademarked name of Everfresh.

Repeated testing of the product has shown it has no adverse effects on the health of test animals. The product works by interfering with the chemical process that causes melanosis. The compound 4-hexylresorcinol is considered a processing aid rather than an additive, even though it is added to the shrimp. Because it is classified as a processing aid, the law does not require it to be listed on a label.

Can Shrimp Be Too Fresh?

Can shrimp be too fresh? Intriguingly, the answer is a qualified yes. Shrimp can never be too fresh to eat. However, they can be too fresh to easily peel. The shells of shrimp that are swiftly washed and iced after catch will stick to their flesh. The solution is to ice them properly (see Freezing Shrimp) and hold them for a day before peeling them. The extra time allows the shrimp's own enzymes to loosen the bond between the flesh and the shell, making peeling easier.

Canning Shrimp at Home

Canned Gulf shrimp used to be a delicacy that even Louisianans, who had lots of fresh shrimp, enjoyed. In the early 1970s, two dozen commercial shrimp canneries existed in Louisiana and Mississippi. By 2005, only one was still in operation and Hurricane Katrina forced it to close.

Shrimp canneries fell by the wayside not just because of shifting consumer tastes, but also because cheaper but vastly inferior canned shrimp imports from Asia flooded the U.S. market. Canned Gulf shrimp were firm and succulent. Their texture and mild taste were good enough to use in almost any dish. A shrimp stew made with canned Gulf shrimp was to die for. Canned imports from Asia are so inferior as to be nearly inedible. Their taste is objectionably strong.

From William D. Chauvin Collection, Louisiana State Archives, Baton Rouge

Even worse is the texture: soft, granular, and crumbly. Cooking with them is impossible.

Those who remember and want to recapture the quality of canned Gulf shrimp may can them at home with a pressure canner. Shrimp, as well as almost all other foods, except tomatoes which are highly acidic, need to be pressure canned to reach temperatures high enough to destroy any stray *Clostridium botulinum* bacteria that could cause botulism food poisoning.

Before canning, shrimp should be deheaded, washed, and dried. Then they should be boiled 8-10 minutes in an acidic brine made with ¾ cup of salt and 1 cup of vinegar per gallon of water. Rinse the boiled shrimp in cold water and peel the tails.

Pack the shrimp into half-pint or pint jars (do not use quart jars), leaving one inch of head space. Cover them with a boiling salt brine made with 3 tablespoons of salt per gallon of water. Wipe the jar rims clean before putting the lid on. Food particles will prevent sealing. Place the jars in the canner and follow the directions in the canner instruction manual exactly.

Some tips:

- Use a pressure canner that is in good condition. Replace the gasket if necessary. Dial pressure gauges should be tested for accuracy at least once a year. Check the internet or write the manufacturer for locations where you can have your gauge tested.
- Re-read and follow directions for the canner. If you no longer have an instruction manual, write the manufacturer for a copy or search the internet.
- Exhaust steam from the canner before closing the petcock or putting on the weighted gauge. Canner directions will specify the venting time required (usually ten minutes).
- Process foods at the correct pressure. At sea level, process foods at 10 pounds of pressure. Increase the pressure ½ pound for each 1,000 feet of altitude above sea level. Very little of Louisiana is more than 400 feet above sea level. Its highest point is Driskill Mountain in Bienville Parish at an elevation of 535 feet.
- To make sure that the pressure stays constant during processing, check the gauge periodically. Weighted gauges should jiggle the number of times per minute specified in canner directions.
- Process foods for the correct length of time. When the required

Photo by Chris Granger

pressure is reached, note the time and continue processing for the specified period. Remove the canner from the heat at the end of the processing time.

- Test seals on jars the day after canning. If the jars have not sealed, use a new lid and reprocess them in the canner for the entire length of time. Alternatively, the product can be refrigerated for use within the next few days or frozen.
- Examine home-canned seafood for spoilage before serving it. Bulging jar lids, spurting liquid, and "off" odor or mold indicate that the food is not safe to eat. Discard spoiled food out of the reach of children and pets. Do not even taste questionable food.

Storing Fresh Shrimp

You've gone to the trouble of locating a premium fresh shrimp producer, you've made the trip down the bayou to get them, and now you don't have time to dehead, wash, pack, and freeze them. What do you do?

Fresh shrimp will last four to five days if properly iced and stored. But they will spoil overnight without proper attention. The amount of ice your shrimp supplier packed with your shrimp is almost certainly not enough. More than likely he packed enough for you to get safely home, assuming that you would handle them immediately.

The first thing to do is to make sure that you have enough ice. As a rule of thumb, use the same volume of ice that you have of shrimp. Ice is not cheap, but shrimp cost more.

Generously layer ice on the bottom of an ice chest, then add some

Courtesy Kim Chauvin, Mariah Jade Shrimp Company

shrimp and top with more ice. Intersperse the shrimp with the ice above them, leaving the ice layer on the bottom undisturbed. Every shrimp should be in contact with ice. Repeat this procedure until all the shrimp are stored. Cap the shrimp with any remaining ice.

Open the drain plug in the ice chest to allow melted ice and any bacteria that it carries to drain. (Obviously, if the shrimp are packed for transport, the drain hole can't be left open. In this case, simply drain the melt water at your first opportunity.)

Place the ice chest in a shady spot where it will receive maximum protection from the sun's rays during the hottest part of the day and in a location where the melt water will not make a mess.

Process the shrimp for use or freezing as soon as possible. Remember that shrimp quality will be at its peak when you first purchase them and that they will only degrade as you hold them unfrozen.

Freezing Shrimp

Shrimp are probably the all-time freezing champs among animal proteins. Properly frozen shrimp will easily last twelve months in the home freezer. If properly frozen, they can be thawed and refrozen twice with no appreciable loss of quality, although some weight loss will occur. They are more resistant to damaging textural changes than are finfish, crabs, and oysters. They do not experience the rancidity problems associated with poultry, pork, or even beef in frozen storage. Their major problem is freezer burn, the loss of moisture from their tissue.

Rapid freezing is important to retain the quality of your shrimp. Rapid freezing produces small ice crystals in the flesh. Slow freezing forms large ice crystals that rupture tissue cells. When the product is thawed, the liquid containing flavor and moisture escapes from damaged cells as thaw loss.

While fresh water freezes at 32°F, the freezing temperature of most seafood is around 25°F, with complete freezing not occurring until 20°F. A freezer should be turned to its coldest setting prior to freezing fresh seafood. It may be returned to a higher setting after freezing is completed, although the lower the storage temperature, the slower that enzymatic changes in the seafood will occur.

If more than a few containers of seafood are to be frozen, they should be spread out in the freezer rather than stacked one on top of the other. Stacking slows the freezing process. After the shrimp are frozen, containers may be stacked.

Maximum quality can be maintained by freezing shrimp in headless form with the shell on the tail. The shell provides added protection against freezer burn. They are best frozen in waxed cartons or plastic containers, although peeled shrimp also can be frozen in heavy plastic bags. No matter the method of storage, after the shrimp are frozen the container should be filled with very cold water and refrozen. This provides an impenetrable ice glaze barrier.

Never add water to a container of fresh shrimp and freeze both shrimp and water at the same time. This slows the freezing process.

Properly Freezing Shrimp

Pack the shrimp snugly in a plastic container.

Snap the lid on the container and place it in the freezer.

After the shrimp are frozen, top off the container with ice water and return it to the freezer

Properly frozen shrimp will be encased in ice.

Thawing Frozen Shrimp

Proper thawing is just as important as proper freezing. Seafood should never be thawed in hot water, or even worse, at room temperature. Thawing at room temperature allows chemical changes and bacterial growth to occur in the thinner or outer portions of the flesh before the center thaws. Thawing in hot water greatly denatures proteins and causes much loss of flavor and moisture.

Thawing seafood under cold running water is the fastest means of thawing. Seafood can also be thawed slowly in the refrigerator, allowing twelve hours for a one-pound package and eighteen hours for a two-pound package. Experts debate which of the two methods is best; however, both are acceptable. For delicate seafood such as fish fillets, thawing in the refrigerator is gentler but demands some advance planning.

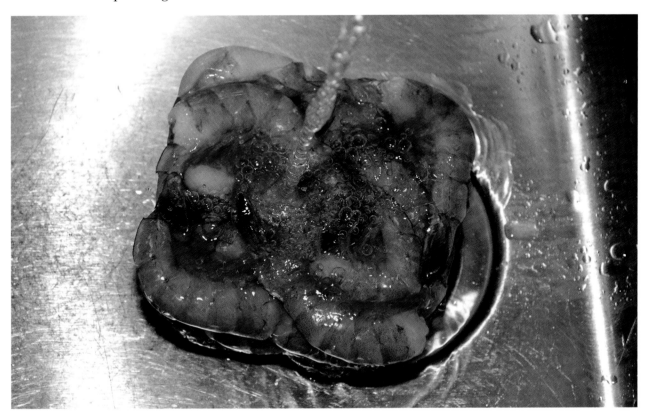

Shrimp and Cholesterol

Shrimp have the reputation of being a food source high in cholesterol. The American Heart Association recommends a diet with no more than 300 milligrams (mg) of cholesterol per day. Shrimp have 150-200 mg of cholesterol for a 3½-oz. serving, compared to 80-95 mg of cholesterol for lean beef, pork, or lamb. Chicken, with or without skin, contains only slightly less.

However, the cholesterol level in human blood appears to be significantly less affected when the dietary source of cholesterol is shrimp or other shellfish than if it comes from other sources. In a peer-reviewed scientific study completed in the 1990s, researchers compared cholesterol levels of people fed diets high in either eggs or shrimp. The researchers measured the changes in the study participants' blood levels of both low-density lipoprotein (LDL) and high-density lipoprotein (HDL). LDL is what we know as "bad cholesterol" because it promotes the formation of plaques that can block arteries and cause heart attacks. HDL, called "good cholesterol," returns bad cholesterol to the liver for removal from the bloodstream.

Those on the shrimp diet did experience a 7 percent rise in LDL levels, but they also had a 12 percent rise in HDL levels. In contrast, those on the egg diet had increased LDL levels of 10 percent and HDL levels increased only 7 percent. In addition, in people who ate the shrimp diet, levels of triglycerides, the form in which fat is carried in the bloodstream, declined by 13 percent.

For unknown reasons, the cholesterol in shrimp is harder to absorb than that from high-fat foods. It is thought that this is because shellfish is lower in total fat and higher in heart-healthy omega-3 fatty acids than meat, eggs, milk, and cheese.

Shellfish Allergies

Food allergies occur when the human body reacts to something eaten as if it were a dangerous substance. To combat the perceived threat, the body's immune system releases antibodies into the bloodstream. These antibodies react with other cells in the blood to cause the release of chemicals, especially histamines.

These processes cause mild to severe symptoms in the allergic individual, depending on how sensitive the person. Less severe reactions such as itchiness, hives and swelling, and sneezing, runny nose, or sinus problems may occur. In a more severe reaction, the throat and tongue can swell and tingling may occur in the mouth. Abdominal pain, diarrhea, and heartburn can be other symptoms, as can wheezing and coughing. A very severe reaction can produce anaphylactic shock, which may be life-threatening. In this state, breathing becomes very difficult and a choking sensation is present. Blood pressure drops and the individual can become unconscious.

Food allergies are common in the United States, affecting an estimated one in twenty-five people. Dairy products and peanuts are common allergies, as is seafood. An estimated 6.5 million Americans have some level of seafood allergy. They seem to be more common among populations that frequently eat seafood than among those who only occasionally eat it.

Seafood allergies fall into three groups or classes: molluscan (oysters, clams, mussels, snails, squid, octopus), crustacean (shrimp, crabs, crawfish, lobsters), and finfish. Some people are able to eat one species in a class but not another. More typically, individuals showing an allergy to a particular seafood will be allergic to others in the same class. Less common is an allergy to all classes or types of seafood. Anyone showing an allergic reaction to any seafood should consult with a doctor before attempting to eat any other seafood.

Susceptibility to seafood allergies seems to be inheritable within families. As a group, African-Americans, at 3.7 percent, are more likely to have seafood allergies than other groups. More women than men experience sensitivity, and more adults than children are allergic to seafood. Children are more likely than adults to have finfish allergies, and crustacean allergies are more common in adults than children.

Often people who develop an allergy to house dust and dust mites

will show an allergy to crustaceans. A reaction can even be triggered in people with a crustacean allergy by aquarium fish food, which often contains brine shrimp or other seafood.

Although thoroughly cooked seafood is less likely to produce a reaction than lightly cooked products, with raw seafood most likely to trigger a serious reaction, the only form of treatment for seafood allergies is to completely remove the allergen from the diet. Severely sensitive people may be advised by a doctor to carry a syringe of epinephrine (adrenaline) with them to counter the effects of accidental exposure, such as that in cooking vapors or cooking oil used for both seafood and other foods.

Fish Oils and Shrimp

All wild-harvested seafood, even extremely lean ones such as shrimp, contain omega-3 fatty acids. These compounds, impossible for the human body to produce, must be consumed in human diets. The ultimate source for omega-3 (often abbreviated n-3) fatty acids lies in the food chain of fish, which begins with phytoplankton, microscopic floating one-celled plants.

The effects of omega-3s were first noted in Greenland Eskimos who ate a diet of fish, whales, and seals. Though their diet was incredibly high in fat as a result of the blubber of whales and seals, they experienced virtually no heart disease. Further research in Japanese fishing villages, followed by a twenty-year study in Holland, cinched the fact that diets high in omega-3s protect against heart attacks.

Thousands of studies in the last twenty years have almost given omega-3s the status of a "magic bullet." Their beneficial effects include:

- Lowering triglyceride levels
- Lowering LDL (bad) cholesterol levels
- Decreasing the occurrence of blood clots
- Lowering blood pressure
- Promoting more flexible blood vessels
- Reducing plaque formation in arteries
- Preventing heart rhythm irregularities
- Easing of rheumatoid arthritis symptoms
- Reducing diseases caused by inflammation (any disease ending in "itis")
- Reducing incidences of breast and colon cancer
- Reducing an array of mental disorders
- Reducing the incidence of migraine headaches
- Promoting the development of the fetal brain and retinas

Achieving the same level of omega-3s as the Eskimos receive from their fish-heavy diet is difficult for most Americans, as the natives of Greenland eat an average of three-quarters of a pound of fatty cold-water fish a day. Higher levels of omega-3s occur in fatty seafood species than in lean species. Coldwater marine species have more omega-3s than warmwater marine species, which in turn are higher in omega-3s than freshwater species.

Louisiana shrimp are lean warmwater creatures. As such, they contain less omega-3s than do fish species such as mackerels, tunas, salmon, and herrings (including sardines), but they can still be an important component in the recommendation to eat two to four seafood meals per week. A not-to-be overlooked benefit of consuming shrimp is that they serve as a lean low-fat substitute for a fattier meat entrée.

From William D. Chauvin Collection, Louisiana State Archives, Baton Rouge

Part II: Recipes

Photo by Chris Granger

The Secret to Cooking Shrimp

The secret to cooking delectable shrimp is simple: don't overcook them! More shrimp dishes are ruined by overcooking than by all other mistakes combined. Small shrimp not overtreated by additives (see Food Additives in Shrimp) should cook through in 2 to 5 minutes. Larger shrimp should need no more than 5 to 10 minutes.

Shrimp do not fall to pieces when overcooked; instead they get tougher, as shrimp release water from their tissues when cooked. In fact, shrimp can lose one-third of their weight to the surrounding dish when properly cooked. Even more is lost if they are overcooked.

This water loss is why it is extremely important to follow instructions in a recipe on whether to use cooked shrimp or raw shrimp as an ingredient. The two are not interchangeable. For example, if raw shrimp tails are used in a recipe calling for 1½ pounds of cooked shrimp, they will add a cup of water to the dish, often with devastating results. The dish can be served watery and diluted, or it can be further cooked to boil the excess liquid out, which will overcook the shrimp and ruin the dish.

Four Star Recipes

Whenever I buy a cookbook, I always wonder which recipe is the best one in the book. If I know or get to meet the author, that is usually the first question I ask. We saved you the trouble in this book. While all the recipes in this book are tested as excellent, we marked what we thought were the "best of the best" as four-star ★★★★ recipes.

Shrimp Dip

There are lots of shrimp dip recipes. Some are better than others, but this is one of the best we've tasted. The cream cheese makes it very rich, but the luscious shrimp taste is undisguised. It's a snap to fix for a get-together and your guests will think that you are a gourmet chef.

1 lb. boiled shrimp, peeled
6 green onions, chopped
1 tsp. lemon juice
½ tsp. hot sauce
½ tsp. Worcestershire sauce
2 cloves garlic, finely minced
1 16-oz. container soft cream cheese
2 tbsp. mayonnaise

Finely chop first 6 ingredients in a food processor. Combine with cream cheese and mayonnaise and mix ingredients together by hand. Serve with your choice of crackers. Serves 4.

Shrimp Spread

Tip: Be aware that chili powder is not just chili powder. Like curry powders, chili powders are blends and every brand is different. All will list chili peppers as the predominant ingredient, but some blends use cayenne peppers, others use ancho peppers, and some use both. Most blends have cumin, Mexican oregano, and garlic in varying amounts. Some have turmeric, cloves, coriander, mace, nutmeg, cinnamon, or black pepper. If you don't like one brand, try others until you settle on a favorite, then stick with it.

This recipe produces a creamy, delicious spread, which when slightly warmed makes a good dip too. The zip from the chili powder, soy sauce, and hot sauce really works for this dish.

1 lb. cooked and peeled shrimp
1 16-oz. container soft cream cheese
4 tsp. prepared horseradish
4 tsp. mayonnaise
2 cloves garlic, very finely minced
⅓ cup very finely minced celery
⅓ cup very finely minced onion
½ tsp. chili powder
2 tsp. soy sauce
1 tsp. hot sauce

Chop the shrimp into very fine pieces. Combine all ingredients and blend well. Chill overnight to allow flavors to blend. Serves 4.

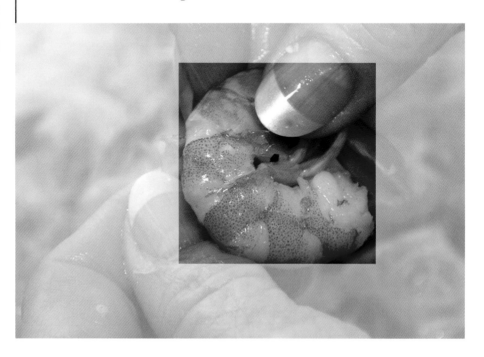

Shrimp Salad

This simple recipe is one of Glenda's favorites. The dish takes 10 minutes to prepare and is a light treat during Louisiana's hot summer days.

2 lb. boiled seasoned shrimp tails
2 boiled eggs
2 stalks celery, diced
2 tbsp. diced green onion
4 tbsp. sweet pickle relish
2 cups mayonnaise

Peel shrimp and eggs and chop finely, either by hand or with a food processor. Mix all ingredients together smoothly. Serve on crackers or toast. Serves 4.

Tip: We like to put leftover boiled shrimp tails in the freezer and use them later for this and other recipes. Leftover boiled seafood of any type—shrimp, crabs, or crawfish—are delightful ingredients for cooking. We have yet to find that the boiling seasonings in the seafood sabotaged a dish.

Shrimp Mold

Tip: Pour-and-boil seafood boil mixes make boiling seafood easy. All the necessary spices are ground and pre-blended with lemon juice crystals and the proper ratio of salt. You simply dump and boil. Several manufacturers produce the blends, and they are retailed in containers ranging from 5-oz. envelopes to 1-gal. plastic jars.

This recipe comes from Claudia Fowler of Baton Rouge, Louisiana. It is absolutely delicious when spread on chips or crackers.

1 5-oz. package pour-and-boil seafood boil mix
1 gal. water
1½ lb. peeled shrimp
2 cups ice
1 10¾-oz. can tomato soup
3 oz. cream cheese
Pinch of baking soda
1 package Knox gelatin
½ cup water
¼ cup chopped celery
¼ cup chopped onion
½ cup chopped bell pepper
1 cup mayonnaise
Salt
Red pepper

Add seafood boil to gal. of water and bring to a boil. Add shrimp and boil until firm, not more than 7 minutes. Remove from heat and add ice. Allow shrimp to soak for 15-20 minutes. Drain, then chop the shrimp finely. Set aside. Heat soup and cream cheese in double boiler until cheese is melted. Add soda. Dissolve gelatin in ½ cup of water. Combine dissolved gelatin with soup mix and allow it to cool. Mix shrimp, celery, onion, bell pepper, and mayonnaise. Add cooled soup mixture to shrimp mixture. Add salt and red pepper to taste. Mold and chill. Serves 8 as an appetizer.

Shrimp and Broccoli Mold

Nova Dee Carbo of Marrero, Louisiana, created this recipe. Like everything she cooks, it is delicious. If you like broccoli, you'll love it with the touch of shrimp. She says that it is an excellent way to use leftover boiled shrimp.

1 10-oz. can Ro*Tel Diced Tomatoes & Green Chilies
1 10½-oz. can cream of mushroom soup
½ cup water, divided
2 8-oz. packages cream cheese
2 envelopes unflavored gelatin
1 lb. boiled shrimp, peeled and chopped
1 cup finely chopped green onions
1 cup finely chopped celery
1 9-oz. box frozen chopped broccoli, boiled
½ tsp. garlic powder

Add Ro*Tel tomatoes and cream of mushroom soup to a saucepan. With ¼ cup of water, rinse the soup can into the pan. Bring to a boil. Lower heat, then add cream cheese and stir. Dissolve gelatin in ¼ cup warm water and add to hot soup mixture. Add shrimp, green onions, celery, broccoli, and garlic powder. Mix well. Bring mixture to a simmer, then pour it into a well-oiled mold and refrigerate overnight. Serve with crackers. Serves 4-6.

Tip: Ro*Tel tomatoes appear so often in south Louisiana cooking that they have almost become an institution. Eight different styles of canned Ro*Tel tomatoes are available. If a recipe calls simply for Ro*Tel tomatoes, use the original version. Ro*Tel Mild and Ro*Tel Hot are also available.

Shrimp au Gratin

This recipe is a 30-year old standby in our kitchen and sets the standard by which all other seafood au gratins are measured. My mother, Hilda, used Velveeta cheese universally for any and all recipes that called for cheese, even lasagnas. When we cast about for just the right cheese to use in this dish, the last cheese we tried was Velveeta, and it was the best. It melts perfectly and complements the subtle taste of the shrimp without being gooey. Serve with garlic bread or toast.

1 lb. peeled shrimp tails
3 tbsp. chopped onion
3 tbsp. melted butter
¼ cup flour
½ tsp. salt
¼ tsp. dry mustard
Dash of pepper
1½ cups milk
1 cup cubed Velveeta cheese
1 tbsp. melted butter
¼ cup dry breadcrumbs

Cut large shrimp in half. Boil the shrimp in water for no more than 10 minutes, until firm and pink. Set aside. Sauté the onion in butter until tender. Blend in flour, salt, mustard, and pepper. Gradually add milk and cook over medium heat until thick, stirring constantly. Add cheese and heat until melted. Stir in the shrimp and pour into a well-greased casserole. Combine the butter and breadcrumbs and sprinkle over the top of the casserole. Bake in 400-degree oven for 10 minutes or until brown. Serves 4-6.

Photo by Chris Granger

Shrimp Rémoulade

This delicious and spicy recipe was contributed by Jennifer Centola. Shrimp rémoulades were one of a dozen or so recipes that appeared on every old-line New Orleans Creole restaurant's menu before Chef Paul Prudhomme and his disciples opened up Creole cooking to exploration. Every restaurant had its own secret rémoulade dressing recipe, but there were enough similarities for each to be recognizable as a rémoulade. Almost all had (and have) horseradish as an ingredient, but this unique recipe does not. It receives its "bite" from the use of shrimp boiled in traditional seafood seasonings, so it is very important to use well-seasoned boiled shrimp to get the right taste. Ketchup, rather than paprika, is used to provide the pinkish color. Like other rémoulades it has green herbs. This one skips the parsley but uses a good amount of celery, which provides a nice crunch and a sweet taste, as well as green onions.

1 tbsp. vinegar
1 tbsp. lemon juice
6 tbsp. olive oil
1 cup Creole mustard
3 tbsp. ketchup
3 green onions, finely chopped
3-4 stalks celery, finely chopped
Salt and pepper to taste
2 lb. seasoned boiled shrimp, peeled
1½ cups shredded iceberg lettuce

Photo by Chris Granger

Combine all ingredients, except shrimp and lettuce. Mix well. Once the sauce is thoroughly mixed, fold in all but 18 of the shrimp. Marinate several hours in the refrigerator. Place shredded lettuce in the bottom of six footed glasses. Top the lettuce with marinade and shrimp. Arrange reserved shrimp around the glass edges. Serves 6 as an appetizer.

Why Creole Mustard?

Creole mustard, a hot, spicy condiment heavily used along the Louisiana and Mississippi Gulf Coast, is different than either prepared yellow mustard or standard whole-grain brown mustards. American prepared mustards are made of the less pungent white mustard seed, which is powdered and then colored bright yellow with turmeric.

In whole-grain mustards, the seed is either left whole, cracked, or coarsely ground, rather than powdered. Creole mustard differs from other whole-grain mustards in the kinds of spices used in it. Horseradish is an important component of Creole mustard. The invention of Creole mustards has been attributed to Louisiana's German Creoles.

Rémoulade Salad Dressing

This salad dressing comes from Gary Percle of Baton Rouge, Louisiana. This excellent rémoulade dressing is most unusual in that it contains neither Creole mustard nor horseradish. The heart of the dish is the Durkee Famous Sauce. Durkee was the very first prepared and packaged salad dressing produced in the U.S. Production began in 1857. Old discarded Durkee bottles have been found along wagon train trails and First Lady Mary Todd Lincoln reportedly served it to Pres. Abraham Lincoln in the White House. Cult followers of the sauce simply call it "Durkee's." This recipe makes a large batch, but it stores well.

1 pt. mayonnaise
1 10-oz. bottle Durkee Famous Sauce
¼ cup olive oil
1 6-oz. jar Dijon mustard
1 cup ketchup
2 tbsp. red wine vinegar
2 tbsp. Worcestershire sauce
2 tbsp. Tabasco sauce
1 tsp. salt
2 tsp. paprika
½ cup chopped green onions
½ cup chopped parsley

Blend mayonnaise and Durkee's with a whisk. Whisk in the oil a little at a time. Add remaining ingredients and mix well. Serve over boiled, peeled shrimp on a bed of shredded lettuce. Garnish with tomato wedges and sliced hard-boiled eggs. Dresses 24 salads.

Shrimp and Mango Spinach Salad

Tip: Do not substitute dried ground ginger for the fresh ginger. It won't work! You will need a zesting tool to prepare the orange zest. Everyone who enjoys cooking needs a zesting tool.

This delightful salad has a touch of sweet, a touch of sour, and a touch of the orient. This recipe goes against our tendency to use ingredients that can be produced locally, but mangos are, oh, so good!

1¼ lb. peeled shrimp
1 lb. flat-leaf spinach
1 large ripe mango
1 medium red onion, thinly sliced
4 tsp. rice vinegar, divided
2 tsp. grated orange rind zest
6 tbsp. orange juice
1 tsp. grated fresh ginger root
¼ cup vegetable oil
1 tbsp. Asian sesame oil
Salt and pepper

Boil, drain, and chill shrimp. Wash and drain spinach thoroughly. Remove stems from spinach leaves and tear into bite-sized pieces. Peel and cut the mango into thin strips. Add mango and shrimp to spinach. Place onion and 2 tbsp. of vinegar in a small bowl. Smash onions in vinegar until onions are pinkish. Add to salad. Whisk together rest of vinegar, orange zest, orange juice, ginger root, vegetable oil, and sesame oil. Add to salad and toss. Salt and pepper to taste. Serve with toast. Serves 4.

Mango Tango

For residents of Louisiana, the nearest source of mangos, a true tropical fruit, is extreme southern Florida. To choose a ripe mango, sniff the stem end. It should have a pleasant perfume-like scent. No smell means no flavor. If it smells sour or like alcohol, it is overripe and has begun to ferment. Ripe mangos have some slight give in their texture when handled. Green mangos will be hard, and overripe ones are quite soft. Unripe mangos may be ripened by being held at room temperature for several days. Mangos do not store well when refrigerated.

Photo by Chris Granger

Shrimp and Pea Salad

Tip: Celery seeds are a good spice for salad dressings, soups, pickles, and breads and there is no such thing as a coleslaw that can't be improved with the addition of celery seeds. They are also delicious when rubbed on large cuts of beef. The principal sources of celery seed are India, China, and France. Indian seeds are darker than the others, have a stronger flavor, and are considered to be the premium seed.

Celery seed adds an herby-nutty element to the taste of this delightful and light salad.

2 lb. shrimp, peeled and deveined
Water to boil
1 10-oz. package frozen green peas
1 red onion, thinly sliced
¾ cup prepared herb, vinegar, and oil salad dressing
½ tsp. celery seed
Lettuce

Add shrimp to boiling water. Return water to a boil, reduce heat, and simmer for 3-4 minutes. Shrimp will be opaque when cooked. Drain and cool immediately. In a large bowl combine shrimp, frozen peas, and onion. In a small container, mix dressing with celery seeds. Stir into shrimp mixture. Serve on lettuce as an appetizer or as a main dish. Peas will be sufficiently thawed by serving time. Serve with crackers or breadsticks. Serves 4-6.

Janice's Shrimp Salad

Janice Lerma of Baton Rouge, Louisiana, secretary for the Louisiana Fisheries Federation, prepared this dish for one of the group's monthly board meetings. The federation was formed in the late 1970s and was Louisiana's lone, near-successful effort to form a professional umbrella group for all types of commercial fishing and seafood processing. Commercial fishermen are so independent and competitive that organizing them has been likened to herding cats. Each will invariably go his separate way, given the first opportunity.

2½ lb. small shrimp, boiled and peeled
1 16-oz. package noodles, boiled
1 cup chopped green onions
1 cup grated cheddar cheese
1 14¾-oz. can drained *petits pois* peas
Mayonnaise to coat
Salt and pepper to taste

Cool shrimp and salad noodles. Mix all ingredients in a large bowl. Gently fold in enough mayonnaise to coat everything evenly, adding salt and pepper to taste. This salad is better if refrigerated for 10-12 hours prior to serving. Garnish with tomato and lemon wedges before serving. Serves 12.

Kickin' Korn Soup

Corn soups are a seafood tradition in south Louisiana. This recipe comes from Larry Roussel, who is one of the better amateur chefs in the region. The only change we made to Larry's recipe was to substitute tasso for andouille because we prefer tasso in all our cooking. This recipe does have some "kick," so if you don't like spicy foods, you might want to use Ro*Tel Mild rather than regular Ro*Tel tomatoes. This has to be the simplest, fastest recipe we have ever used, and like everything from Larry, it is delicious.

1 lb. small peeled shrimp
1 10-oz. bag frozen chopped onions, bell peppers, and celery
3 11-oz. cans Mexicorn
1 10¾-oz. can cream of shrimp soup
1 6-oz. can tomato sauce
1 10-oz. can Ro*Tel Diced Tomatoes & Green Chilies
3 cups water
1 lb. tasso, cut into ½-inch pieces
3 tbsp. flour
1 tsp. salt
1 tsp. black pepper
1 tsp. garlic powder
1 heaping tbsp. dried parsley flakes
1 tbsp. Kitchen Bouquet

Blend all ingredients together in a large pot. Bring soup to boil over high heat, stirring often. Reduce heat until soup boils at a slow roll. Cook 10 minutes, stirring occasionally. Serves 5-6.

Tasso & Andouille

Both tasso and andouille are made of pork and used as seasoning meats. But there the resemblance ends. Tasso (pictured in right of photograph) is very lean, highly seasoned pieces of pork, typically ham, which is heavily smoked. Its use originated in the Cajun communities west of the Atchafalaya Basin.

Andouille is a large, highly seasoned, heavily smoked pork sausage. The pork in andouille has been cut into chunks rather than ground. It is best known as being from the German Coast of Cajun Country, the parishes straddling the Mississippi River north of New Orleans. A variation of eastern Louisiana andouille is also produced in a small area west of the Atchafalaya Basin.

Marie Hymel's Corn Soup with Shrimp

Glenda loves corn soups. When she ate this one at Hymel's Restaurant in Convent, Louisiana, she declared it simply the best. Thanks to Marie Hymel for sharing her restaurant's recipe with us. Hymel's Restaurant has been open since the 1940s and for most of that time, the shrimp they used were freshwater river shrimp from the Mississippi River. Few fishermen fish for river shrimp anymore, so Hymel's has had to increasingly resort to using saltwater shrimp.

1½ lb. peeled shrimp
3 tbsp. plain flour
⅓ cup vegetable oil
2 medium onions, finely chopped
2 stalks celery, finely chopped
1 14¾-oz. can cream-style corn
1 14¾-oz. can whole-kernel corn
1 14½-oz. can diced tomatoes, undrained
3 qt. water
Salt and pepper to taste
1 large bell pepper, chopped
2 tbsp. chopped parsley

Rinse shrimp and set aside. To make the roux, combine the flour and oil in a large, heavy pot (preferably cast iron) over high heat. Stir constantly to keep from burning, until roux becomes golden brown. Add peeled shrimp and cook about 10 minutes. Add onions, celery, corn, and tomatoes and simmer for 1½ hours over low heat. Add water and continue to simmer for at least another 1½ hours. Add salt, pepper, bell pepper, and parsley and cook at least another ½ hour. If the soup is too thick, add more water and bring to a boil. Cook until it has the consistency of a vegetable soup. Serves 6.

Courtesy Byron Despaux

Shrimp and Corn Soup

This is a great recipe from the kitchen of Jewel Boudreaux of Lafitte, Louisiana. Ordinarily, corn soups are not one of my favorite seafood dishes, but when I ate this one, I had to have the recipe. It is one you won't be disappointed with.

Jewel is an extraordinary cook. She has the uncanny ability to eat a dish then mentally deconstruct it to its individual ingredients. Then she goes home and cooks it, adding her own touches. Jewel also produces a knockout shrimp gumbo (she never would give me the recipe).

1 stick butter
2 tbsp. flour
1 large onion, chopped
¼ cup chopped green onion
¼ cup grated provolone cheese
1 qt. milk
2 14¾-oz. cans whole kernel corn
2 14¾-oz. cans cream-style corn
1 10½-oz. can cream of mushroom soup
1½ lb. small peeled shrimp
½ tsp. Worcestershire sauce
4 bay leaves
Salt and cayenne pepper to taste

Photo by Chris Granger

In a large pot, melt butter over low heat, then blend flour into the melted butter. Add onions and sauté until wilted. Add remaining ingredients and cook over medium heat for 40 minutes. Stir frequently to prevent scorching. Serves 8.

Shrimp and Mushroom Soup

This is a recipe for mushroom lovers. We like mushrooms and find that a ratio of a pound of shrimp to a pound of mushrooms gives a nice balance.

1 lb. shrimp tails
1 lb. fresh mushrooms, sliced
1 stick butter
3 cups chicken broth or 3 bouillon cubes dissolved in 3 cups water
1 tsp. dry dill
Salt and white pepper to taste
¼ cup flour
½ pt. cream

Peel the shrimp. If the shrimp are large, cut them in halves or thirds, then set aside. Sauté sliced mushrooms in butter. Add broth to mushrooms and simmer. Stir dill, salt, and white pepper into the broth. Mix flour with a little water, then stir in a little at a time until soup thickens as desired. Cook an additional 5-10 minutes. Add shrimp and heat until cooked. Add cream and serve hot. Serves 4.

Experimenting with Mushrooms

Don't be afraid to experiment in your cooking with varieties of mushrooms besides the common white *Agaricus* mushroom. Of the 38,000 or so known species of wild mushrooms, about 2,500 are grown agriculturally. Before I began gathering wild mushrooms, an experienced mushroom-lover, or mycophile, as he liked to call himself, told me that different species of mushrooms taste as different from each other as different vegetables do—comparing green peas to sweet corn for example. He was right.

Unquestionably, one of the best exotic mushrooms is the chanterelle, the most common species of which is *Cantharellus cibarius*. This trumpet-shaped orange mushroom is extremely aromatic and full flavored. It is a common mushroom, especially during autumn, in Louisiana Florida Parish hardwood uplands, and fortunately, it is also farmed.

The oyster mushroom, *Pleurotus ostreatus,* is another common wild mushroom in Louisiana that is also farm raised. It is a semi-translucent white in color and grows on tree trunks and stumps in lowlands and swamps. It seems to especially favor willow trees. When the oyster mushroom is found in hill country, it is usually growing on a decaying magnolia tree log or stump. It has a mild taste and a dense texture, so it is often thinly sliced before cooking. Oyster mushrooms are most commonly found during the winter months.

Unless you are in the hands of an experienced mushroom hunter, it is best not to gather wild mushrooms because of the risk of poisoning. However, enough exotic species are available in grocery stores to sat-

isfy your curiosity for a while. Besides chanterelle and oyster mushrooms, farmed crimini, shiitake, enoki, portabella, and occasionally porcino and morel mushrooms are available locally.

Zuppa Gambero

We named this dish in honor of the strong Italian contribution to New Orleans Creole cooking. It simply means "shrimp soup."

¼ lb. butter
1 tbsp. minced garlic
¼ cup chopped green onions
1 4-oz. can sliced mushrooms
½ cup chopped red bell pepper
¼ cup chopped green bell pepper
1 lb. peeled medium shrimp
2 tbsp. flour
1 oz. white wine
2½ cups chicken stock
1 tsp. lemon juice
1 tsp. chopped parsley
Cornstarch, optional
Salt and pepper to taste

In a heavy pot, melt butter over medium heat. Add garlic, green onion, and mushrooms and sauté for about 2 minutes. Add bell peppers and sauté until peppers soften. Add shrimp and sauté just long enough for shrimp to turn pink. Do not overcook them. Sprinkle in flour and stir until any liquid in the pan is absorbed. Stirring, add wine then chicken stock. Increase heat to medium-high. Mix in lemon juice and parsley. Cook until the sauce thickens. If the sauce boils too long before thickening, the shrimp will overcook, so add cornstarch if necessary. Salt and pepper to taste. Serves 4.

Choosing Cast-Iron Cookware

Whenever a recipe recommends using a heavy pot or frying pan, Glenda and I both reach for one of our cast-iron pieces. Yes, we have the omnipresent aluminum and stainless cookware; yes, we have top-of-the-line Calphalon pans; and yes, we have an authentic carbon steel wok, but we really love to cook with cast iron. We also collect it. All the other stuff gets used (or we would garage sale it), but cooking traditional foods in cast iron gives us a connection to our ancestors. The food tastes better too.

Part of the appeal might be the fact that some iron does indeed migrate from the pot into the food. The amount of increased iron in food can be as small as 28 percent for quick-cooking, relatively low-moisture corn bread to 46 percent for acidic spaghetti sauces on up to a 1,400 percent increase in iron in a medium white sauce.

At one time dozens of American companies manufactured cast-iron cookware. The "mad rush to Magnalite" has bankrupted all but one, Lodge Manufacturing. We find Lodge cast iron to be a little disappointing, as its inside cooking surfaces have a rough, grainy texture that can be difficult to clean, even in well-seasoned pieces. Some imports are even worse, having deep concentric rings that look like they were machined into the metal. Our absolute favorite brand is Griswold, made from 1865 to 1957. Old, but perfect pieces of both Griswold and that other American cast-iron icon, Wagner, can be found at antique stores, auctions, garage sales, and "junque shoppes." Look for pieces with perfectly smooth cooking surfaces. Even a mild amount of pitting is unacceptable. A few pieces of good cast-iron cookware should be part of every serious cook's *batterie de cuisine.*

Paw Paw's Shrimp and Bean Soup

This recipe comes from Ervin Hebert, who was a commercial crab fisherman from Westwego, Louisiana, when he gave the recipe to us.

½ lb. pickled pork
¼ cup vegetable oil
2 medium onions, diced
3 green onions, diced
2 stalks celery, diced
1 medium bell pepper, diced
1 lb. small peeled shrimp
2 qt. water, divided
1 8-oz. can tomato sauce
3 cups cooked white beans
6 oz. #4 spaghetti
Salt and pepper to taste

Dice pickled pork into ½-inch cubes and fry in vegetable oil in an 8-qt. pot until golden brown. Sauté onions, green onions, celery, and bell pepper with the pork over medium-low heat for 10 minutes. Add shrimp and ½ cup water and simmer 5 more minutes. Stir in tomato sauce and cook 30-40 minutes. Add beans and remaining water and bring to a boil. Add uncooked spaghetti and cook until done. Season with salt and pepper to taste. Serves 8.

Tip: The white beans referred to in this recipe are navy beans. Small navy beans are usually preferred over large navy beans. If you don't know how to cook down dried beans, simply substitute Blue Runner cream-style navy beans.

Bean Soups

Bean soups are a favorite Cajun dish in cold weather months. While some white beans are eaten in the city and a fair amount of red beans (kidney beans) are eaten in the country, red beans are far and away the choice in New Orleans Creole cooking and white beans are more popular in Cajun country, especially in that part east of the Atchafalaya Basin. There, heaven can be defined as white beans and rice with fried catfish or smothered rabbit.

Louisiana bean soups differ from plain white beans by the addition of tomato sauce or stewed tomatoes and water. Some bean soups use spaghetti, but most bean soups call for small amounts of rice.

Shrimp and Asparagus Soup

Asparagus is a delightful vegetable, but it refuses to grow in the Deep South. By the time we get it in our stores, the bottom ends of the stalks have usually become woody and tough. Glenda and I usually cut off this tough end then gently peel the lower inch or two of each remaining stalk with a potato peeler. It's a lot of work, but good asparagus is worth the effort.

1½ lb. fresh asparagus
¼ cup margarine
⅓ cup flour
1 tbsp. salt
¼ tsp. pepper
Dash of nutmeg
1½ qt. milk
1 lb. small peeled shrimp
3 cups grated sharp cheddar cheese
Paprika

Clean asparagus and cut into ½-inch pieces. Boil until tender, then drain and set aside. Melt margarine in a large pot. Blend in flour, salt, pepper, and nutmeg. Add milk gradually, stirring constantly until thickened and smooth. Add asparagus, shrimp, and cheese. Cook over low heat until cheese melts and shrimp are cooked. Garnish with paprika. Serves 6.

Pepper Stew

This recipe comes from Mrs. M. J. Ford of Harvey, Louisiana. (She insists that this is how her name is to appear in print, without her first name.) She says that this dish, which has a pronounced bell pepper taste, is New Orleans soul food. The use of sweet bell peppers, as opposed to hot peppers, appears in all sorts of south Louisiana cooking, although not to the extent to which they are used in this recipe. Bell peppers are part of the "holy trinity" of seasonings, along with onions and celery, which is a cornerstone of modern Cajun cooking.

1 large onion, chopped
4 cloves garlic, minced
4 large sprigs parsley, chopped
2 tbsp. shortening
4 large bay leaves
1 15-oz. can tomato sauce
3 green bell peppers, diced
3 red bell peppers, diced
3 lb. shrimp tails, peeled
Salt and pepper to taste

Sauté the onion, garlic, and parsley in shortening until tender. Add bay leaves and tomato sauce and simmer for 10 minutes. If the sauce is too thick, add a little water, but be careful not to add too much because juice from the shrimp and peppers will help thin out the sauce. Stir in the green and red bell peppers. Simmer 5 more minutes. Add peeled shrimp and boil about 5 minutes. Season with salt and pepper to taste. Serves 6.

Suzy's Shrimp Chili

How about this one? When Suzy Despaux of Barataria told me about her recipe for shrimp chili, I was intrigued. I envisioned something dark, loaded with potent chili powder and maybe some added adobo seasoning. Instead, what I got was a pleasantly balanced, aromatic dish. It tastes like no other seafood dish we have ever eaten. We think you will love it.

3 tbsp. olive oil
½ stick margarine
4 cloves garlic, minced
2 large bell peppers, chopped
2 medium onions, chopped
3 stalks celery, chopped
2 green onions, chopped
1 tsp. parsley flakes
1 tsp. salt
½ tsp. pepper
2 6-oz. cans tomato paste
2 14½-oz. cans chopped tomatoes
1 10¾-oz. can tomato soup
3 tbsp. Worcestershire sauce
¼ tsp. Italian seasoning
1 tsp. chili powder
½ tsp. sugar
1 tsp. filé
1 tsp. sweet basil
2½ lb. boiled shrimp, finely chopped
½ lb. very small raw whole shrimp tails
Cooked rice or pasta, optional

In oil and margarine, sauté the garlic, bell pepper, onion, celery, green onion, and basil until tender. Add salt and pepper and mix. Stir in the tomatoes and seasonings and simmer over a low heat until the gravy is smooth, at least 30 minutes. Add the shrimp and cook an additional 5-7 minutes, until the whole shrimp are done. Serve by itself or over rice or pasta. Serves 8.

Shrimp Boulettes

This recipe comes from Lois and Mike Oliver of Estay Fish Company in Grand Isle, Louisiana. It's simple to prepare and just delicious.

Boulettes are small balls formed of ground meat or seafood mixed with seasonings, and they often use egg for a binder. They may or may not be breaded before frying. They can be made small for hors d'oeuvres or for use in bisques or made larger and flattened for use in sandwiches. Boulettes have come to be almost as associated with Cajun/Creole cooking as gumbo.

2 lb. headless shrimp
1 clove garlic, minced
1 medium onion, chopped
½ bunch green onions, chopped
2 eggs
1 tbsp. flour
Salt and red pepper to taste
Cooking oil

Peel the shrimp and chop the tails. Combine shrimp with garlic, onions, green onions, eggs, flour, and salt and red pepper. Mix thoroughly and shape into balls about an inch in diameter. Deep-fry balls in cooking oil until golden brown. When done, the boulettes will float to the top of the oil. Serves 4.

Shrimp and Broccoli Braid

When I first saw Terrebonne Parish resident Jessica Daigle's preparation, I thought it was a braided king cake, that New Orleans Mardi Gras season tradition. When I checked closer, I saw that it contained seafood. When I tasted it, I loved it. The only hitch to this delightful dish is that you will need a baking stone. But what the heck, you probably wanted one anyway. Note that this dish takes two days to prepare.

½ tsp. black pepper
½ tsp. ground cumin
½ tsp. oregano
½ tsp. garlic powder
2 tbsp. soy sauce
Juice of 1 medium lemon
1 lb. peeled shrimp, washed and drained
1 cup chopped broccoli
½ cup chopped bell pepper
1 cup shredded sharp cheddar cheese
1 clove garlic, minced
½ cup mayonnaise
2 8-oz. packages crescent rolls
1 lightly beaten egg white
2 tbsp. slivered almonds

In a mixing bowl, combine pepper, cumin, oregano, garlic powder, soy sauce, and lemon juice. Add shrimp. Mix well. Cover and marinate in the refrigerator overnight. Remove shrimp from refrigerator and chop into small pieces. Sauté the shrimp in the marinade sauce in a saucepan over medium heat until the sauce is creamy. Remove from heat. Combine broccoli, bell pepper, cheese, garlic, mayonnaise, and chopped shrimp in a mixing bowl. Mix well. Preheat oven to 375 degrees. Unroll 1 package of crescent rolls. Do not separate into individual rolls. Arrange the longest side of dough across the width of a 12 x 15-inch baking stone. Repeat with the second package. Using a rolling pin, roll dough to edges of baking stone, sealing the perforations. Along the longest side, cut dough into strips 3 inches deep and 1½ inches apart. Repeat on opposite side. Place filling in the middle of the dough and spread evenly along its length. Braid dough strips over filling. Brush with egg white. Sprinkle with almonds. Bake 25-30 minutes. Serves 4-6.

Making Shrimp and Broccoli Braid

Open and unroll two cans of crescent dinner rolls. Do not separate them. Place the rolls on a baking stone side by side.

Roll the dough with a rolling pin to seal the perforations.

On the long sides of the dough, cut strips 1½ inches apart and 3 inches deep.

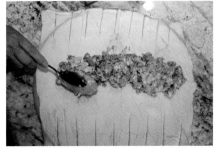

Spread the filling along the center of the dough.

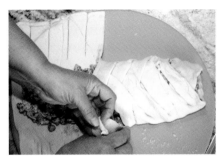

Braid the dough strips over the filling.

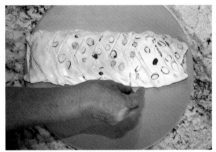

Brush the top of the braid with egg whites and sprinkle with almonds.

Jerald's Favorite Pickled Shrimp

Tip: You may feel a powerful urge to substitute a more trendy vinegar such as white wine, balsamic, or rice vinegar for the distilled vinegar called for here. Don't! They won't work as well. The clear sharpness of distilled white vinegar is needed. Surprisingly, because of the sugar, the end result is a mild but delicious taste.

For the onion in this recipe, I particularly favor that hard-to-find, spicy, red Louisiana Creole onion. It is never available in grocery stores. Small family fruit stands in Lafourche and Terrebonne Parishes are where Creole onions are most available.

This dish should be eaten within 2 or 3 days of preparation.

This is a simple recipe, and the celery seed and capers really set it off. Don't leave them out or you will loose the dish.

2 lb. peeled shrimp tails
2 medium onions, peeled and sliced into very thin rings
1½ cups vegetable oil
1½ cups white vinegar
½ cup sugar
1½ tsp. salt
1½ tsp. celery seed
5 tbsp. capers with juice

Boil shrimp 3-5 minutes in salted water. Drain, rinse with cold water, and chill in ice water. Once shrimp are chilled, remove from ice water. Alternate layers of shrimp and onion rings in a sealable container. Mix remaining ingredients and pour over the shrimp and onions. Seal and place in refrigerator for 6 hours or more, shaking occasionally. Remove shrimp from marinade before serving. Serves 6.

Pickled Shrimp and Peppers

Unfortunately, pickled and smoked seafood do not play a major role in Louisiana cuisine. However, shrimp lend themselves particularly well to pickling.

This is a good summertime treat. While most vegetable gardens begin to quail before Louisiana's summertime heat, bell peppers begin to copiously ripen to their mature red, yellow, or orange colors and reach their peak of sweetness.

2 lb. peeled shrimp
2 medium red onions, sliced into thin rings
½ cup cubed green bell pepper
½ cup cubed red bell pepper
1½ cups vegetable oil
1½ cups white vinegar
½ cup sugar
1½ tsp. salt
1½ tsp. celery seed
4 whole cloves garlic
2 tbsp. capers with juice

Place shrimp in boiling, salted water and simmer 3-5 minutes until pink but still tender. Drain and rinse the shrimp. Alternate layers of shrimp, onion rings, and pepper in a sealable container. Combine the remaining ingredients and pour over the shrimp. Close the container and put in the refrigerator for 12 hours or more. Serves 6 generously.

Photo by Chris Granger

Marinated Shrimp

Tip: The key to preparing shrimp is not to overcook them. When boiling shrimp, heat the water first. Add the shrimp, bring the water back to a boil, and boil for 2-5 minutes, no more. When the shrimp are pink and opaque but still tender, they are done. Always rinse in cool water or drop them in ice water for a few minutes to prevent them from continuing to cook from residual heat.

Shrimp lend themselves particularly well to marinades. The idea for this recipe came from chatting with a bunch of Alabama shrimpers during a break at a shrimpers' meeting.

4 lb. peeled medium shrimp
4 envelopes of Italian salad dressing mix
¼ cup balsamic vinegar
½ cup water
2 cups olive oil
1 can black olives, drained and sliced
1 13-oz. jar salad olives
4 stalks celery, thinly sliced
1 red onion, thinly sliced
8 green onions, chopped
10 cloves garlic, minced

Boil shrimp, drain, and set aside to cool. Mix salad dressing with vinegar, water, and olive oil in a large bowl. Add cooled shrimp. Add remaining ingredients, cover, and marinate overnight in a refrigerator. Use within 2 days or the shrimp will toughen. Makes 12 servings.

Shrimp à la Cantrelle

Simple is good! Sometimes we forget that. When Phil and Bernice Cantrelle of Jennings, Louisiana, sent us this recipe, we thought that it was too simple, as we were accustomed to long lists of ingredients and lots of steps. However, good south Louisiana cooking is supposed to be simple.

Seasoning Mix
1 cup salt
⅔ cup red pepper
⅓ cup garlic powder
¼ cup Accent
¼ cup white pepper

Combine all ingredients and mix together thoroughly. Store in airtight container.

1 stick butter
1 lb. headless shrimp
Seasoning mix to taste

Melt butter in a large skillet. Season the shrimp with seasoning mix and add them to the skillet. Stir constantly while cooking over medium heat for 3-5 minutes or until shell slightly separates from meat. Do not overcook. Overcooking will toughen the shrimp and cause the shell to stick to the meat. Sample a shrimp for seasoning. If the salt is to your taste, the dish is ready to serve. Serves 4.

Tip: You may substitute the Cajun/Creole seasoning of your choice if you choose not to use the Cantrelle seasoning. If you use the Cantrelle mix, you will have a generous amount left for other uses. Generic monosodium glutamate may be substituted for Accent.

Shrimply Delicious

This very good dish is one easily adapted to Louisiana tastes, as it is well-seasoned without being overly spicy and served over rice. Cajuns west of the Atchafalaya Basin center much of their cuisine around smoked and specialty meats. Cajuns east of the Basin have a more heavily seafood-influenced diet. One thing that both subcultures share, however, is a love of rice. So important is rice that many are perfectly happy to eat steamed rice unadorned by anything but plain butter. New Orleans city cooking, often called Creole cooking, interchangeably uses pasta, especially spaghetti, with rice. This recipe will also work with pasta.

¼ cup flour
1 clove garlic, minced
¾ tsp. salt
¼ tsp. pepper
1 lb. headless shrimp, peeled
¼ cup cooking oil
1 16-oz. can stewed tomatoes
1 4-oz. can chopped green chili peppers
¼ cup white wine
Cooked rice

Combine flour, garlic, salt, and pepper. Roll the shrimp in the seasoned flour then fry in oil until lightly browned. Add tomatoes, chili peppers, and wine to the shrimp. Heat thoroughly, stirring occasionally. Serve over cooked rice. Serves 4.

Shrimp Fontenot

The basics for this recipe came from Beth Fontenot, who grew up in Franklin, Louisiana, but has globe-trotted to Washington, Texas, and even Africa. Beth shares with her mother, Ginger, also an impressive cook, a love of saffron rice. She says saffron and plain long-grain rice pair equally well with this dish. The Heinz 57 Sauce is an ingredient we had never used in any other seafood dish. Beth has orders to bring back original west African gumbo recipes for us when she returns.

2 lb. peeled shrimp
½ cup chopped green onions
½ cup chopped green bell pepper
½ cup chopped yellow bell pepper
½ cup chopped celery
4 tsp. chopped parsley
1 4-oz. can sliced mushrooms
½ tsp. unflavored meat tenderizer
6 oz. beer
5 tbsp. prepared mustard
½ tsp. cayenne pepper
½ tsp. seafood seasoning
1 tsp. garlic powder
⅓ cup Heinz 57 Sauce
5 dashes soy sauce
Cooked rice

Mix all ingredients, except rice, and stir well. Marinate in refrigerator at least 2 hours or overnight. Cook mixture in large skillet over medium heat, stirring frequently, until sauce is reduced to the desired thickness and shrimp are done. Serve over rice. Serves 4.

Roatan Shrimp

This unusual recipe comes from Tammy Macaluso of St. Bernard Parish, Louisiana. It is named after the Caribbean Island of Roatan, one of the Bay Islands off the coast of Honduras. The Bay Islands consist of 3 major islands, Roatan, Utila, and Guanaja, as well as other smaller ones. While the Bay Islands are now part of Spanish-speaking Honduras, during colonial times they were mostly under the control of Great Britain, and today, English is the language spoken there. Some Bay Islanders cannot speak Spanish at all.

The largest of the islands, Roatan, is the most heavily developed for tourism. Utila, the smallest island, and Guanaja are less developed. In fact, Guanaja has no roads. The Bay Islands are known for excellent diving opportunities and great seafood.

The Bay Islands have historically had a close connection to the port city of New Orleans, with large communities of Bay Islanders living here while the menfolk worked as seamen on merchant ships. After retiring, many moved back to the islands. Their best culinary secret is the treasured mutton pepper, used in so many of their recipes.

This recipe has a distinctive tropical flavor. In fact, the dish is very sweet. If you are a lover of the taste of the tropics and shrimp, you'll like this one.

2 tbsp. vegetable oil
½ cup chopped green pepper
½ cup chopped onion
1½ lb. raw peeled shrimp
2 tbsp. flour
1 15-oz. can cream of coconut
1 10¾-oz. can cream of shrimp soup
¼ tsp. cayenne pepper
Salt to taste
½ tsp. paprika
Cooked rice

Heat the oil in a heavy 10-inch skillet or Dutch oven. Sauté the green peppers and onions until tender. Add shrimp and cook over a low fire until pink. Do not overcook. Add flour and stir. Add remaining ingredients, except rice. Simmer over low fire for 20 minutes, stirring occasionally. Serve over rice. Serves 4-6.

Courtesy Louisiana Seafood Promotion and Marketing Board

Jambalaya à la Dee

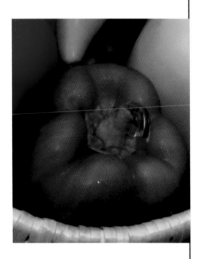

Nova Dee Carbo, called Dee, produces some of the best recipes. They are all originals, not found in any other cookbook. Dee lives in Marrero, Louisiana, and worked with me for several years in the early 1980s.

This recipe is for a red jambalaya, one that contains tomatoes. This Creole jambalaya from the New Orleans area contrasts with rural Cajun jambalayas, including those from Gonzales, the home of the jambalaya festival and cooking contest, which are brown in color and contain no tomatoes. Needless to say, each side thinks theirs is the best and the only "real" jambalaya.

1 medium onion, chopped
¼ cup chopped bell pepper
1 tbsp. bacon drippings
3 thinly sliced yellow squash
1 lb. peeled shrimp
½ tsp. salt
½ tsp. pepper
½ tsp. garlic powder
1 14½-oz. can stewed tomatoes
1 cup uncooked long-grain rice

Sauté the onion and bell pepper in bacon drippings until the onion is transparent. Add squash and simmer until tender. Add shrimp, salt, pepper, garlic powder, and stewed tomatoes and simmer until shrimp are pink. Mix in the rice, cover, and cook over low fire for approximately 20 minutes or until rice is tender. Serves 4.

Smothered Shrimp

Contrary to what our friends "up north" think, Cajun cooking is not about using exotic or weird ingredients (unless crawfish is considered weird) that are overly seasoned with red pepper. Smothering, the process of cooking with low heat, moisture, and onions, is a basic Cajun cooking technique—nothing flashy, just delicious. It is most often used with vegetable dishes such as snap beans, okra, or potatoes but it also works really well with shrimp. The simple list of ingredients in this dish will allow the taste of the shrimp to come through undisguised, so start with the right ingredient—fresh, local shrimp.

4 large onions, slivered
⅓ cup cooking oil
2½ tsp. salt
2 tsp. black pepper
1 tsp. garlic powder
1 cup water
2 lb. medium shrimp tails, peeled
Cooked rice

In a heavy pan, sauté the onions in the oil over medium heat. The onion slices should end up wilted and clear with toasted edges. Add salt, pepper, garlic powder, and water. Bring to a boil and add shrimp. Cook over medium-low heat until shrimp are done (no longer translucent). Do not overcook or the shrimp will become tough. Serve over cooked rice. Serves 4.

Do Onions Make You Want to Cry?

In spite of the importance of onions in Louisiana cuisine, for many people peeling and cutting onions is a tear-jerking experience. As onions are cut, the knife naturally damages the bulb's cells and causes chemical reactions that produce syn-propanethial-S-oxide, a gas that forms sulfuric acid when it reaches the eye. There, it irritates the iris and pupil, the colored parts of the eye, causing tear glands to produce tears in order to flush out the acid.

Contact lens wearers seldom experience as much irritation as others because the lens covers the iris and pupil. Besides wearing contact lenses, there are a few things that can be done to reduce onion-induced eye irritation. Refrigerating onions before processing them will significantly reduce irritation by slowing down the reaction that produces the irritating gas. Using a sharp knife helps a lot too. It will slice cleanly and through fewer cells; a dull knife mangles and macerates a wide path, damaging more cells. It is also recommended that cooks either not cut off the root end of the onion or cut it last, as it has a higher concentration of enzymes. Finally, working near a lit candle or stove burner will help somewhat, as the flame will burn some of the gases.

Smothered Shrimp and Cucumbers

When Barbara Olano of Barataria, Louisiana, told me to pull up a chair and try some of this dish, I thought she was kidding. Who ever heard of cooking cucumbers with shrimp? After I tasted it, I became a believer. It is one of the best seafood casserole dishes I have ever eaten. This is Barbara's invention and not found in any other cookbook. Her husband, Shelby, a lifelong commercial shrimper, said the idea came from having to find a use for those large yellow overly mature cucumbers that always manage to hide themselves in the vines until they grow too large for fresh use. This is a wonderful example of ingenuity and originality in south Louisiana cooks.

½ cup vegetable oil
2½ lb. peeled, cubed, and seeded overmature cucumbers
1½ lb. peeled shrimp
1 large onion, chopped
1 tsp. Kitchen Bouquet
Salt and pepper to taste
½ cup seasoned breadcrumbs

Heat the vegetable oil in a skillet. Fry cucumbers, shrimp, and onions in oil over medium heat until the water is reduced. Add Kitchen Bouquet for color and salt and pepper to taste. Mix in breadcrumbs and cook 5 minutes. The cucumbers will make their own water, so do not add any. Serves 6.

Shrimp-Stuffed Mirliton

Even though it comes from Delanea Crochet of Baton Rouge, Louisiana, this recipe is an old New Orleans favorite. The dish was a gold medal winner in the Louisiana Seafood Promotion and Marketing Board-sponsored 4-H Seafood Cookery Contest. I served as a judge for the contest and was quite impressed by the dish.

4 fresh mirlitons
4 tbsp. margarine
1½ cups finely chopped onion
½ cup finely chopped celery
½ cup chopped parsley
1¼ cup seasoned breadcrumbs, divided
Salt and pepper to taste
2 lb. peeled shrimp, chopped

Boil the mirlitons in water until they are tender when pierced with a fork. Remove from water and let cool. Cut the mirlitons in half and use a spoon to scoop out the pulp into a mixing bowl. Discard the seeds but save the skins. Arrange the 8 halves on a baking sheet. In a saucepan over medium heat, melt the margarine and sauté the onions, celery, and parsley until the onions are transparent. Stir in the mirliton pulp, 1 cup breadcrumbs, salt, and pepper and mix well. Reduce the heat to low and cook for 10 minutes. Add chopped shrimp; stir and remove from heat. When cool enough to handle, spoon the shrimp mixture into the mirliton shells. Sprinkle the tops with the remaining breadcrumbs. Bake in preheated oven at 350 degrees for 30-40 minutes. Serves 4.

Mirliton Preparation

Place the mirlitons in a pot with enough water to cover them. Boil until tender.

Remove the mirlitons from the boiling water and allow them to cool. Slice them in half lengthwise and remove the seed.

Scoop the flesh from the mirliton, retaining ⅛ inch of flesh with the peel to form a shell for stuffing.

The Mirliton

Before the working-class New Orleans neighborhoods of the Irish Channel and downtown broke up in the 1970s, the stuffed mirliton (inevitably pronounced "melaton") was practically a signature dish of those areas. It was often stuffed with ground beef, but shrimp-stuffed mirliton was especially relished. While this member of the squash family was so popular in the city, it was seldom eaten by Cajuns west of Des Allemands.

Part of its popularity in New Orleans was due to the tiny house lots of the city, with their postage-stamp-sized backyards, which had no room for a garden. A mirliton plant could be shoehorned into the tiniest bare spot and the plant would be allowed to ramble with its vines up to 50 feet long over the one-car garage, fence, and even house. Once established (they can be hard to sprout) the plant seemed to thrive on neglect in the subtropical climate of New Orleans.

The mirliton, more properly known as chayote but also called vegetable pear, is a native of Mexico that has spread through most of Latin America and across the world to Africa, many parts of Asia, Australia, Italy, Norway, Portugal, France, Russia, and Slovenia. In the United States, however, it is only commonly found in the Southwest with its large Mexican-heritage population, southern Florida, and New Orleans. How it came to be part of Creole cuisine is undocumented, but not surprising. Until the rise of Miami and, to a lesser degree, Houston as Hispanic portals, the Port of New Orleans served for 200 years as the gateway to Latin America and the Caribbean.

During winter's freezes, the mirliton plant dies back to the ground, sprouting again in spring. It grows vigorously until late summer and then blooms. Single pale-green, pear-shaped fruits up to 8 inches long are ready to pick 30 days later. One good plant can produce more fruit than a large family can eat.

Unlike its squash, melon, and cucumber relatives with their many seeds, each mirliton fruit contains a single 1- to 2-inch seed (which is also edible, as are the plant's root, stem, and leaves). The pale, firm to crisp flesh of the mirliton is extremely bland and benefits from aggressive seasoning during preparation.

Glenda's Shrimp-Stuffed Potatoes

Potatoes have not played a central role in the cooking of south Louisiana, even though everyone eats them. You will not commonly find in Cajun/Creole cooking the potato gratins, puffs, soups, casseroles, pancakes, gnocchi, and knishes of the Midwestern and Northeastern United States. Seldom eaten in south Louisiana are boiled potatoes and gravy, and baked potatoes are only eaten upon occasion, usually with beef steak. Besides smothered potatoes, they have only secured a place in Cajun cuisine as that ubiquitous warm, semi-mashed potato salad always added to a bowl of gumbo. Nevertheless, Glenda loves twice-baked potatoes. This is her creation.

4 large baking potatoes
½ cup margarine
½ cup half-and-half
¼ cup chopped green onions
1 cup grated sharp cheddar cheese
¾ tsp. salt
½ tsp. pepper
1 lb. cooked peeled small shrimp
Paprika
2 tbsp. chopped parsley

Bake potatoes for 45 minutes at 400 degrees. When done, remove from oven and allow to cool. When cool to touch, cut potatoes in half lengthwise. Scoop out the pulp, leaving a firm shell about ¼ inch thick. Combine potato pulp, margarine, half-and-half, green onions, cheese, salt, and pepper. Whip until smooth. Mix in shrimp. Stuff the potato shells with shrimp mixture and sprinkle with paprika. Bake at 425 degrees for 10 minutes. Garnish with parsley. Serves 4.

Shrimp-Stuffed Squash

This recipe came to us from Celie Robin of Yscloskey, Louisiana. Celie and her late husband, Charles, were shrimpers for many years. Both are Isleños, descendents of settlers from the Canary Islands who came to Louisiana between 1778 and 1783. Celie's shrimp cooking was featured by the *New Orleans Times-Picayune* in 1998.

Until very recently, only two types of squash were used in rural south Louisiana kitchens, the white pattypan summer squash and the large cushaw. Cushaw were cooked sweet, typically baked, and the pattypan was cooked savory. In south Louisiana the word "squash" always referred to pattypans, not zucchini, yellow crooknecks, or butternuts.

4 white pattypan squash
1 cup chopped onions
1 tbsp. olive oil
1 cup ham, roughly ground in a food processor
1 cup cooked peeled shrimp, roughly ground in a food processor
1½ cups Italian breadcrumbs, divided
2 tbsp. butter
1 egg

Microwave whole squash on high for 28 minutes or boil in water until tender. Cool squash and cut the tops off, then scoop out the pulp into a bowl, retaining shells for stuffing. In a large skillet, sauté onions in olive oil. Add ham, shrimp, and squash pulp. Cook for 10 minutes. Add the breadcrumbs, retaining a few tablespoons, and the butter and stir in the egg. Stuff the mixture into the shells. Sprinkle with remaining breadcrumbs. Bake in 350-degree oven until brown, about 45 minutes. Serves 4.

Isleños in Louisiana

The Isleños (Islanders) of St. Bernard Parish are part of Louisiana's wonderful cultural mélange and have long been among the most important seafood suppliers to the New Orleans area. They are descendents of Spanish-speaking settlers from the Canary Islands, located about 100 miles west of the Moroccan coast of Africa.

When Spain acquired Louisiana in 1762, the Spanish governor, Bernardo de Gálvez, recognized that the colony needed people, especially soldiers, to settle the land and fend off interlopers. Seven hundred Canarian men were recruited in 1777-78 from 5 of the 7 Canary Islands. Including their families, 2,373 people left for Louisiana on 7 vessels.

Four Canarian settlements were formed: San Bernardo de Gálvez in modern-day St. Bernard Parish, Valenzuela near the junction of Bayou Lafourche and the Mississippi River, Gálveztown on the Amite River near Manchac, and Barataria at the site of the modern village of Barataria. All were founded in 1779.

The Barataria settlement had the shortest life. It was the smallest community and was flooded by Mississippi River crevasse waters and 2 hurricanes in 2 years. Many Canarians left by 1782 and the settlement was considered abandoned by 1785. Most went to St. Bernard, although some stayed and took up fishing as a lifestyle. Many people with Spanish surnames live in the area today, but most identify their name as Filipino rather than Canarian.

The Gálveztown settlement was only slightly less unfortunate. Located in a low area, it suffered from repeated flooding and rampant disease outbreaks. The Canarian settlers rapidly began to move from the site. Some went to Baton Rouge and settled in what is now known as Spanish Town. Others stayed closer, moving to higher ground nearer the Mississippi River. This community became known as Galvez and then later as Gonzales. The original settlement site was deserted by 1807. Although the Canarian culture and identity were absorbed by those of later settlers, many Spanish-surnamed families are still present.

Valenzuela was more successful, having been founded on higher ground. Although the settlement was stable, the Canarians assimilated themselves with the more

Courtesy Susan Robin

numerous Acadian settlers. Vestiges of Canarian culture hung on until the 20th century in the various brulees, isolated settlements on bayou banks in the low country away from Bayou Lafourche, and Spanish surnames are very common in the area today.

The Canarian settlement in St. Bernard is the only one of the four to have maintained its Spanish identity to the present, probably because it remained relatively isolated from the more numerous French- and English-speaking communities of Louisiana. They are known today as Isleños (pronounced ees-LANE-yos). Founded on the ridges of Bayou Terre aux Boeuf, the early settlement was known by several names, including Nueva Gálvez and Concepción, before becoming San Bernardo de Galvez or St. Bernard. The settlement, augmented by some Spanish additions, became self-sufficient by the mid-1780s.

While it was founded as an agricultural settlement and indeed supplied New Orleans with truck farming produce, many Isleños gravitated to the life of fur trappers and commercial fishermen. In the isolated trapping and fishing communities of Delacroix (the Island), Reggio, Yscloskey, Shell Beach, and Hopedale, an archaic Louisiana-Spanish dialect was preserved, as was (barely) the décima, a 10-line a cappella style of folk song.

With the exception of the Isleños, Canarians were so thoroughly acculturated into the greater southeast Louisiana community that little of their culinary influence (jambalaya may or may not be an exception) is identifiable. The most notable Canarian dish still prepared is the Caldo of St. Bernard Parish. Caldo is a soup made with salt meat or pickled meat and a wide array of vegetables.

Many Isleño families relocated from St. Bernard Parish following the massive destruction caused by Hurricane Katrina. Still, the Los Isleños Festival is held each spring at the Isleños Museum Complex in St. Bernard. Many Isleños travel from new homes in St. Tammany Parish and other places to the marshes of St. Bernard Parish to follow the fishing lifestyle.

Courtesy Cecile Robin

Courtesy Patsy Assevado

Shrimp-Stuffed Peppers

It is indeed unfortunate that many people pick the stuffing out of stuffed bell peppers, eat it, and throw the peppers away rather than cut them up and eat them with the stuffing. Green bell peppers, like any unripe fruit, are more pungent than the ripe fruit. Try cooking this dish with ripe peppers instead of green ones. Depending on the variety, ripe peppers are red, yellow, orange, or occasionally even purple or brown. They are much sweeter and milder than green peppers and as an added benefit are even higher in vitamin C and beta carotene than green peppers. Your guests will thank you.

1 tsp. salt
3 cloves garlic, slivered
2 qt. water
6 medium bell peppers
1 10½-oz. can cream of mushroom soup
Juice of 1 lemon
Dash of black pepper
2 tbsp. grated onion
2 tbsp. + 6 pats butter, divided
½ cup cooked rice
1 lb. peeled, cooked shrimp tails
1 cup grated cheddar cheese
Paprika

Place salt and garlic in pan with water. Bring to a boil. Cut tops off peppers, scoop out center, and boil for 10 minutes. Remove peppers and discard the water. Combine soup, lemon juice, pepper, onion, and 2 tbsp. butter in a saucepan. Cook over low heat until butter melts. Add rice and shrimp to sauce. Mix well. Stuff peppers with mixture and place upright in a baking dish. Top with cheese, pat of butter, and paprika. Add ½ inch of water to dish. Bake at 350 degrees for 30 minutes. Serves 4-6.

Photo by Chris Granger

The Pepper Story

Peppers are a mainstay ingredient in south Louisiana cooking. Bell peppers are one leg of the holy trinity of Cajun seasonings, along with onions and celery. In addition to bell peppers, ground cayenne pepper is an important cooking spice and various bottled hot sauces are used both in cooking and at the table.

Peppers, members of the genus *Capsicum*, are native to Central and South America. Seeds have been uncovered in Mexican excavations from 7,000 years ago. Although these may have been wild peppers, it is thought that peppers were cultivated there at least 2,000 years ago. They, along with allspice, were brought back to Spain in 1493 by Christopher Columbus. From the Iberian Peninsula they spread through Europe and into India and Indonesia. Portuguese explorers carried pepper seeds to West Africa. Hot pepper use revolutionized cuisines and became so ingrained in much of the cooking of Asia and Africa it is easy to assume that peppers were always used there.

Peppers derive their "heat" from capsaicin, a compound that has no flavor or odor. It works its magic (some would say damage) by stimulating pain receptors in the mouth and nerve endings in the skin. Capsaicin also stimulates the production of saliva and gastric juices, which serves to increase appetite and aid in digestion.

The piquancy of peppers is measured in Scoville heat units (SHU), which indicate the amount of capsaicin present in the pepper. A solution of the extract of the pepper is diluted in sugar water until a panel of fine tasters can no longer taste any heat. For example, a pepper with an SHU rating of 5,000 would have to be diluted by 5,000 times as much sugar water to be undetectable. Obviously, this method has some flaws, as human judgment, no matter how expert, is involved. Even if the judgment were exact, variations can exist within a pepper variety due to climate, soil type, and individual plant variations.

Observers have long noted that the mildest peppers are more popular in countries with cooler, temperate climates such as the United States and Europe, and hotter peppers are favored in hotter countries. Bell peppers, which have a Scoville heat unit rating of less than 100, are heavily used in the U.S., including Louisiana. They are almost never eaten in Mexico, in spite of the large amounts of them grown there for export to the U.S.

Researchers have found 22 species of wild pepper and 5 species of domesticated pepper. The bell pepper, as well as two other peppers used in Louisiana cooking, the cayenne and the jalapeno, belongs to the species *Capsicum annuum*. Along with the bell pepper, the cayenne and jalapeno are both considered relatively mild. The jalapeno has a Scoville rating of 2,500-10,000 and the cayenne's rating is 30,000-50,000. Other *C. annuum* pepper varieties include the pepperoncini (0-500 SHU) and the serrano (8,000-22,000 SHU). Another pepper popular in Louisiana is the tabasco, which belongs to the species *C. frutescens*. Like the cayenne, it has an SHU of 30,000-50,000.

The hottest peppers in the world belong to the species *C. chinense*. Included in this species are the habenero (100,000-300,000 SHU), the scotch bonnet, used in Jamaican cooking (150,000-325,000 SHU), and the absolutely incendiary naga jalokia from northeast India at 855,000-1,000,000 SHU.

Spanish Shrimp

Tip: Most confusing to non-Louisianans trying to use recipes produced here is the tendency for Louisianans to use the terms "green onions" and "shallots" interchangeably. The two are definitely not the same. Almost always, unless the recipe is from a sophisticated New Orleans chef, what we intend to be used are green onions, because we want the taste of the tops. True shallots are an entirely different member of the onion family and have a garlic-onion taste. Real shallots can be hard to find and a little pricey.

This is a concoction that I put together while trying to come up with a shrimp stuffing for bell peppers and it's really quite good as a casserole. The recipe contains the holy trinity of Cajun cooking—onions, bell pepper, and celery—as well as two other staple seasonings, garlic and green onions.

1 medium onion, chopped
1 bell pepper, chopped
1 large stalk celery, finely chopped
3 cloves garlic, chopped
2 tbsp. cooking oil
1 10-oz. can stewed tomatoes with chilies
1 10-oz. can stewed tomatoes
½ cup chopped green onions
½ cup chopped parsley
2½ lb. small headless shrimp, peeled
1½ cups cooked long-grain rice
Salt and red pepper
Seasoned breadcrumbs

In a large skillet or saucepan, sauté onion, bell pepper, celery, and garlic in oil until soft. Stir in tomatoes, green onions, and parsley and cook until heated thoroughly. Add shrimp and cook uncovered over medium heat until liquid cooks away or shrimp are done. If shrimp are done and liquid remains, spoon off excess liquid and discard. Mix shrimp mixture with cooked rice in a casserole dish. Season with salt and pepper to taste and top with breadcrumbs. Bake in 350-degree oven for 10 minutes or until breadcrumbs brown. Serves 6.

Celery-Shrimp Casserole

Don't let all the celery in this recipe scare you. It blends quite well with the shrimp and is not overpowering. Celery, a member of the holy trinity of Cajun cooking, often gets a bad rap as being bitter and stringy. In discussing cooking we often hear people commenting that they don't want too much celery in a dish. They probably remember the dark green tough stuff that someone smeared processed cheese on years ago and tried to get them to eat raw. The varieties available now are better and growers do a better job of blanching the stalks. With today's celery, even the outer stalks in a bunch are light and tender. Celery also completely changes character when cooked, becoming sweet and aromatic.

⅓ cup chopped celery
⅓ cup chopped green pepper
⅓ cup chopped onion
3 tbsp. margarine
1 lb. shrimp, peeled
1 10¾-oz. can condensed cream of celery soup
⅓ cup canned sliced water chestnuts
1 hard-boiled egg, chopped
1 tbsp. lemon juice
¼ tsp. salt
¼ tsp. pepper
⅔ cup cracker crumbs
½ cup shredded cheddar cheese

In a 2-qt. saucepan, sauté celery, green pepper, and onions in margarine for 10 minutes. Stir in all of the remaining ingredients, except for the cheese, and spoon the mixture into a 1½-qt. casserole dish. Bake in 350-degree oven for 15 minutes. Sprinkle cheese on top and return to oven until melted. Serves 4.

Shrimp and Eggplant Casserole I

Tip: Powdered bay leaves provide a more direct flavor boost but can be difficult to find. One source is Penzey's Spices at www. penzeys.com or 1-800-741-7787.

This recipe contains one of my favorite herbs, bay leaves. I love bay leaves (Glenda says that they are tasteless) but don't use them as often in my seafood cooking as I do in tomato sauce dishes and with venison, waterfowl, smothered vegetables, and beef roasts. The fragrance of the bay leaf in cooked dishes is stronger than the taste and is produced by the oil eugenol.

1½ lb. peeled shrimp tails
1½ medium eggplants, peeled and cubed
½ cup chopped onion
½ cup chopped parsley
¼ cup chopped green pepper
2 cloves chopped garlic
⅓ cup olive oil
1 16-oz. can tomatoes, drained
2 bay leaves
1 tsp. salt
½ tsp. pepper
½ tsp. thyme
⅔ cup water
3 tbsp. melted butter
¾ cup breadcrumbs

Boil shrimp then set aside to drain. In a saucepan sauté onion, parsley, green pepper, and garlic in oil until tender but not brown. Add eggplant, drained tomatoes, bay leaves, salt, pepper, thyme, and water to mixture. Cover and simmer until eggplant is tender. Stir in shrimp. Place mixture in a greased casserole dish. Combine butter and breadcrumbs and sprinkle crumbs evenly on top of mixture. Bake uncovered at 400 degrees for 35-40 minutes. Serves 6.

Bay Leaves

Bay leaves are not used heavily in Cajun or Creole cooking, except for boiling seafood. This is surprising, as the use of bay leaves is a fixture in Mediterranean dishes, and Louisiana cuisine has strong Mediterranean influences. Bay leaves are used in such classic French preparations as bouillabaisse and bouillon.

Occasionally, one will find a rural south Louisianian who still picks and uses leaves from the red bay, *Persea borbonia*, a south Louisiana freshwater swamp tree. Its 3- to 6-inch leaves are quite a bit larger than the 1- to 2-inch leaves of the Mediterranean bay, the classical bay leaf. Mediterranean bay is a large evergreen bush that grows well in south, and even central Louisiana, especially if it is planted in a somewhat sheltered location.

Shrimp and Eggplant Casserole II

Tip: Be sure to use fresh, ripe tomatoes, not the canned product. They add a distinct, but not overpowering taste.

It is important not to tinker with the ingredients. The fluids in them are necessary to perfectly cook the rice.

In our search for interesting recipes, we occasionally bump into one that is hard to describe. Seafood/eggplant dishes are common, but this one is a little different. The rice in the recipe pleasantly dominates the eggplant and, while this is not a spicy dish, the ingredients are so well balanced that even a "pepperhead" will not be tempted to sauce it at the table.

1 large eggplant
2 tbsp. salt
1 medium onion, finely chopped
2 cloves garlic, minced
½ cup olive oil
2 tomatoes, peeled and chopped
¾ cup uncooked rice
1 lb. small peeled shrimp
¼ cup water
1 tsp. salt
¼ tsp. pepper
¼ tsp. basil
1 cup grated mozzarella cheese

Peel eggplant and cut into 1-inch cubes. Place eggplant and salt in a large bowl with just enough water to cover the eggplant. Top the vegetable with a heavy bowl or plate to keep the eggplant immersed. Soak for 20 minutes then drain thoroughly. Sauté onion and garlic in olive oil. Add eggplant and sauté until the outside edges are transparent. Add tomatoes, rice, shrimp, water, and seasonings. Cover and simmer for 10 minutes or until shrimp are just pink. Pour into a greased 2-qt. casserole and top with mozzarella. Cover and bake at 350 degrees for 30 minutes. Serves 4.

Shrimp Petit Pois

This is a tasty, easy-to-cook casserole that combines two favorites, shrimp and mushrooms.

1½ lb. shrimp tails
1 lb. fresh mushrooms
¼ cup melted butter
⅓ cup flour
½ tsp. salt
½ tsp. red pepper
Dash paprika
1 qt. milk
1 10-oz. package frozen peas, thawed
Breadcrumbs

Boil and peel the shrimp; set aside. Wash and slice the mushrooms and sauté in butter. Remove the mushrooms from the pan and set aside. Blending well, make a sauce of the butter, flour, salt, pepper, paprika, and milk. Pour sauce into a casserole dish and mix in mushrooms, shrimp, and thawed peas. Cover generously with breadcrumbs and bake in a 350-degree oven for 10 minutes. Serves 4.

Tip: Be sure to thaw the peas or they won't cook through.

Shrimp 'n' Shrooms

Shrimp and mushrooms go well together. This recipe marries the two tastes perfectly. It's also quick and easy to cook.

4 tbsp. chopped bell pepper
4 tbsp. chopped onion
2 4-oz. cans mushrooms
4 tbsp. margarine
2 10½-oz. cans cream of mushroom soup
1 cup grated cheddar cheese
½ tsp. pepper
2 lb. peeled shrimp
½ cup dry breadcrumbs
2 tbsp. melted margarine

Sauté bell pepper, onion, and mushrooms in margarine until tender. Stir in mushroom soup, cheese, and pepper. Heat, stirring constantly, until cheese melts. Add shrimp. Pour mixture into a well-greased 1-qt. casserole dish. Combine breadcrumbs and melted margarine. Sprinkle over top of shrimp mixture. Bake in 400-degree oven for 10 minutes or until thoroughly heated and crumbs are brown. Serves 4.

Shrimp Louisiane Casserole

Casseroles seem as American as apple pie. However, it appears that the word "casserole" was coined in France around 1600 and came from the word *casse*, a roomy, round crock used for slow cooking. In the United States, casseroles reached a peak of popularity between 1960 and 1980. During this time, more wives began working out of the home, resulting in families with limited cooking time. Because casseroles are so convenient, their popularity blossomed. Ingredients can be put together quickly and popped in the oven for their flavors to blend together unattended.

2 slices white bread
½ cup milk
1 cup chopped onion
¾ cup chopped bell pepper
1 clove garlic, minced
2 tbsp. margarine
1 lb. peeled shrimp
1 10½-oz. can cream of mushroom soup
3 cups rice, cooked
1 tbsp. chopped parsley
1½ tbsp. lemon juice
1½ tsp. salt
¼ tsp. black pepper
¼ tsp. red pepper
Paprika

Soak bread in milk. In a saucepan cook onion, bell pepper, and garlic in margarine until softened. Add shrimp and continue cooking 3 minutes longer. Stir in soup, rice, parsley, lemon juice, salt, and black and red pepper. Add soaked bread and mix well. Spoon the mixture into a buttered shallow 2-qt. casserole dish. Sprinkle with paprika. Bake covered at 350 degrees for 30 minutes. Serves 4.

Shrimp Pie

Seafood pies are another of those preparations common in Louisiana cooking that have no clearly identifiable origin. Using good fresh mushrooms is important for this dish.

1 cup sliced mushrooms
¼ cup chopped green onions
6 tbsp. butter, melted
4 eggs, well-beaten
1½ cups half-and-half
1 tsp. salt
1 tsp. pepper
⅛ tsp. dry mustard
⅛ tsp. nutmeg
1 cup shredded mozzarella cheese
1 lb. small peeled cooked shrimp
1 unbaked 9-inch pie shell

In a small saucepan, sauté mushrooms and green onions in butter until tender. In a large bowl combine eggs, half-and-half, salt, pepper, dry mustard, nutmeg, and cheese. Mix well. Stir in shrimp, mushrooms, and green onions. Pour into pie shell. Bake at 400 degrees for 15 minutes. Reduce heat to 300 degrees and continue to bake for 40 minutes. Serves 4.

Mushroom Management

When purchasing fresh *Agaricus* (white) mushrooms, always select those that are uniformly creamy white in color. Darker mushrooms or those that appear bruised are going downhill and have a short shelf life. Older mushrooms will develop a stronger and more intense mushroom taste, which is not bad in itself but can become overwhelming.

Agaricus mushrooms, unlike some other varieties, are pretty sturdy and can be cleaned by rinsing them in the sink or in a colander. Stubborn dirt particles can be picked off by hand or gently wiped. After rinsing, they should be allowed to drain or be patted dry with paper towels. It is best not to expose mushrooms to water more than an hour before cooking.

Before storing mushrooms, remove them from any air-tight packaging. Instead, store them in a brown paper bag in the refrigerator. Good fresh mushrooms can last a week, but mushrooms are prone to drying out and absorbing flavors from their surroundings. To prevent drying, put the paper bag of mushrooms inside a perforated plastic bag.

Skinny Shrimp Cupcakes

Betty A. Cortez of Des Allemands, who contributed this recipe, says it is based on one from the 1990 Weight Watchers food program. When we first looked at it, we wondered, "How can it be good with no salt or pepper?" Then we tried it and we loved it. It really features the marvelous taste of shrimp.

12 oz. peeled cooked shrimp, crumbled
2 tbsp. ranch dressing
2 tbsp. cooking oil
4 slices bread, crumbled
1 egg
12 cupcake papers

Combine all ingredients and mix thoroughly. Divide evenly into 12 cupcake papers. Bake at 350 degrees for 10-15 minutes until brown. Serves 4 dieters or 2 hearty eaters.

Shrimp Manicotti

This is a true-blue invention of Gay Matherne, formerly of Barataria, Louisiana. It is so good, I can hurt myself eating too much of it. This is one of those dishes that you will serve to guests whom you want to impress. Our daughter Lisa has claimed this recipe as one of hers, but she still bugs Glenda to cook it for her when she visits.

2½ lb. medium shrimp tails
½ cup cooking oil
1 box manicotti noodles
1 lb. cubed mozzarella cheese
1 lb. cubed longhorn-style colby cheese
½ cup Parmesan cheese
½ cup Italian breadcrumbs
1 tbsp. garlic powder
1 tbsp. dried parsley
2 eggs
Salt and pepper
1 16-oz. jar spaghetti sauce

Photo by Chris Granger

Peel and rinse shrimp. Heat water to a boil in a large pot. Once the water begins to boil, add the cooking oil. Drop noodles into boiling water one at a time. Boil until done, drain, and rinse well. Mix together the remaining ingredients, except the spaghetti sauce, and carefully stuff into the noodles. Place the noodles in a single layer in a glass baking dish and cover with the spaghetti sauce. Cover and bake at 350 degrees for 35-40 minutes. Serves 6.

Elmer's Island Angel Hair

This recipe comes from our good friend Sandy Corkern of Franklinton, Louisiana. He cooked it for the two of us while we camped on the beach one evening after a day of surf fishing on Elmer's Island. This recipe is simple yet delicious. Before you add garlic or any other ingredients, try it like it is.

1 12-oz. package fresh angel hair pasta
1 stick margarine
2 tsp. liquid crab boil
1½ tsp. seasoned salt
1½ lb. peeled shrimp, cut in half if desired
1 tbsp. chopped parsley

Cook pasta according to package directions (approximately 1-2 minutes), drain, and return to the pot. While pasta is cooking, melt the margarine in a saucepan, then add crab boil and seasoned salt. Sauté the shrimp in the margarine mixture but do not overcook them. Remove shrimp from saucepan and pour margarine mixture over drained pasta, tossing to coat. Add shrimp and parsley and toss to mix. Serves 4.

Shrimp and Tasso Pasta

Pat Attaway of Lafayette, Louisiana, gave this delightful recipe to us. Pat is a dedicated Atchafalaya delta duck hunter and also loves to fish for speckled trout in Vermillion Bay in the late summer and early fall. He is a master at cooking ducks on the bar-b-que and the smells emanating from his duck hunting houseboat when he is at the grill can make your knees buckle. He is also addicted to that prairie Cajun treat, Hebert's stuffed boneless chickens.

For this dish, he says that a pound of crawfish tail meat or 2 dozen oysters may be substituted for the shrimp. However, if oysters are used, the cream sauce will need to be cooked slightly longer because the liquid lost from the oysters will thin the sauce. Pat also says that for a delicious variation, a can of drained artichoke hearts may be added. We use shrimp without artichoke hearts and find it delicious.

1 pt. heavy cream
½ lb. tasso, diced to ¼-inch cubes
¾ tsp. salt
¼ tsp. black pepper
¼ tsp. red pepper
½ tsp. dried basil
½ tsp. dried thyme
1 lb. spaghetti
1 lb. peeled shrimp
½ cup chopped green onions
½ cup chopped parsley
Parmesan cheese (optional)

Pour the cream into a large heavy skillet over medium heat. Stir the cream when it begins to rise to keep it from overflowing. When it comes to a boil, immediately add the tasso, salt, black and red pepper, and herbs. Simmer for 8-10 minutes or until the cream sauce becomes thick. You can prepare the sauce ahead to this point. While the sauce simmers, boil the pasta in water until al dente, then drain. Stir into the sauce the shrimp, green onions, and parsley and simmer until the shrimp turn pink, about 3-4 minutes. Ladle the sauce over the pasta and toss. Serve with freshly grated Parmesan cheese if desired. Serves 4-6.

Shrimp and Spaghetti

I got this recipe from Bev Lopez. For many years, she and her husband, Ed, owned and operated Ed and Bev's Seafood on the corner of Transcontinental Drive and Airline Highway in Metairie, Louisiana. This is a "clear" gravy based around olive oil and butter.

2 lb. medium to large shrimp, peeled
2 sticks butter
¼ cup olive oil
Juice of 2 lemons
3 cloves garlic, crushed
6 green onions, chopped
½ cup sauterne wine
Tony Chachere's Original Creole Seasoning to taste
Cooked spaghetti
French bread

If the shrimp are large, butterfly them. Melt butter in a small pan and mix in the olive oil and lemon juice. Pour the butter sauce into a 9 x 13-inch banking dish. Add garlic, green onions, and wine and stir to blend. Place shrimp in sauce and sprinkle Creole seasoning over them. Place in broiler for 5 minutes. Turn shrimp, sprinkle with more Creole seasoning, and broil 5 more minutes. Serve over boiled spaghetti. Use French bread to sop up the juice. Serves 4-6.

Pick Your Gravy

One of the major ways in which Creole (New Orleans) cooking differs from Cajun cooking is in the types of sauces, usually called "gravies," that are used in each. Cajuns make roux-based gravies as well as onion gravies, which are centered around slowly cooking down lots of slivered onions in a little cooking oil. They seldom make a red gravy, except for basic ground-meat spaghettis. White gravy, made with unbrowned flour, is seldom prepared in Cajun country.

Creole cooking uses some onion gravies, lots of brown and white gravies, and red and clear gravies. Both red and clear gravies, called "sauces" by sophisticates, clearly owe their origin to the strong Italian (usually Sicilian) influence on Creole cuisine.

Shrimp Pasta

This delicious dish is classic New Orleans-style Creole cooking showcasing the Italian influence on Creole cuisine. The recipe was donated by Wayne Schexnayder, who was a chef at the old French Market Restaurant in New Orleans. Schexnayder, who owns Schexnayder's Acadian Foods, says the recipe is equally good for oysters.

Tip: We use 1 lb. of fresh fettuccine when preparing this dish.

1 cup olive oil
1 cup (2 sticks) butter
Juice from 1 lemon, retain peeling
2 tbsp. granulated garlic
1 tsp. black pepper
4 tsp. Italian seasoning with rosemary
1 tbsp. Worcestershire sauce
4 oz. white wine
1 lb. medium shrimp tails or oysters
1 cup mushroom stems and pieces
Cooked pasta

Combine all ingredients except shrimp, mushrooms, and pasta in a saucepan. Cook over medium-high heat, stirring constantly for 4-6 minutes or until all liquid evaporates, leaving only the olive oil. Add mushrooms and shrimp and cook until shrimp turn pink (or the edges of the oysters curl). Serve over cooked pasta. Serves 4.

The Italian Contribution

Cooking in the New Orleans area has been shaped by strong influences from France, Acadia, Spain, Germany, the Canary Islands, Africa, Italy, the West Indies, Croatia, and Latin America. The latest major group to overlay its culture is the Italians, or more properly the Sicilians. While some Italians have lived in New Orleans since its early days, major immigration occurred between 1889 and 1910. Some of the early immigrants were from northern and central Italy, but Sicilians rapidly came to dominate the influx.

Some Italians settled in rural south Louisiana, but most ended up in New Orleans. Their impact on cooking and culture was important, but their impact on the seafood industry was profound. Almost all started out as laborers, but through strong work ethic and perseverance, many built small businesses. By 1975, more than three-fourths of the 120 wholesale and retail seafood businesses in the metropolitan New Orleans area were owned by descendants of Sicilian immigrants.

Noodly Shrimp

This is a dish we whipped up using some leftover noodles (Glenda and I both hate to waste leftovers). We started by adding some shrimp and then that ubiquitous cream of mushroom soup. One thing led to another, and then, voila! It's good enough to cook often.

1 10½-oz. can cream of mushroom soup
¾ cup milk
1 lb. peeled cooked shrimp
2 cups cooked egg noodles
½ cup sour cream
3 tbsp. chopped parsley
½ tsp. dried dill
½ tsp. salt
¼ tsp. pepper
¼ cup dry breadcrumbs
1 tbsp. melted margarine

Mix soup and milk in a large bowl. Add shrimp, noodles, sour cream, parsley, dill, salt, and pepper. Mix well. Pour into a well-greased 8 x 8-inch baking dish. Combine breadcrumbs and margarine and sprinkle over the top. Bake at 350 degrees for 20-25 minutes. Serves 4.

Shrimp iMonelli

This recipe was developed by Brian Blanchard, owner and chef of iMonelli Restaurant in Lafayette, Louisiana. iMonelli opened in 1983 and hired Brian as dishwasher and bus boy in 1984. He worked his way to waiter, then manager, and finally executive chef before buying the restaurant in 1988. Shrimp iMonelli is his creation and has been on the menu since 1989. Brian describes iMonelli as having south Louisiana cuisine with Italian influence. Brian also owns À La Carte Cafe and Catering in Lafayette and Café Jo Jo's in Morgan City.

1 lb. large shrimp
1½ tbsp. olive oil
1½ tsp. nutmeg
1½ tsp. salt
1½ tsp. black pepper
1½ tsp. crushed oregano
1½ tsp. minced garlic
⅓ cup white wine
⅓ cup heavy cream
3 tbsp. fresh Parmesan cheese
Pasta of your choice

Peel the shrimp and sauté them in olive oil for 4-5 minutes or until they turn pink. Add nutmeg, salt, black pepper, oregano, garlic, and wine. Bring to a boil; boil 2 minutes then remove from heat. Remove shrimp from pan. Add cream and Parmesan cheese to the sauce. Bring to a boil. Add shrimp and simmer about 5 minutes, until thickened. Serve over your favorite pasta. Serves 4.

Tip: For this recipe, be sure to use the freshest Parmesan cheese possible. If it is dried out, it will make the sauce "gritty."

Parmesan Cheeses

Not all Parmesan cheeses are created equal. There is Parmesan cheese, there is Grana Padano cheese, and there is Parmigiano-Reggiano cheese. The latter is produced in the Parma and Reggio Emilia areas of Emilia-Romagna in northern Italy. Gourmets consider this cheese to be the only "true" Parmesan cheese.

Grana Padano is very similar to Parmigiano-Reggiano and is produced in Italy, but in Lombardy. It is only aged 15 months, compared to 2 years for Parmigiano-Reggiano, and the milk used to produce it contains slightly less fat.

Another cheese appearing more often is Reggianito. This cheese is made in Argentina using methods imported from northern Italy by dairy farmers. Aged a year or less, it is much less expensive than either Parmigiano-Reggiano or Grana Padano. It is also substantially less flavorful.

The most well known of the generic Parmesan cheeses is Kraft's American version of Parmesan. It is aged only 10 months, the curds in production are cut into bigger pieces that don't drain moisture as well, and it contains much more salt than Parmigiano-Reggiano. It is also sold grated, which some experts say causes it to lose much freshness. American Parmesan is usually viewed by cheese experts as much inferior in flavor and texture.

Brown Shrimp Creole

Other than the presumption that shrimp Creole originated in New Orleans, it is difficult to untangle the origins and history of this dish. Shrimp Creoles are typically red, or as they say in New Orleans, "They have a red gravy." All manner of processed tomato products can be used—paste, sauce, canned, stewed, etc.—depending upon the preferences of the cook. Another characteristic of shrimp Creoles is that enough bell pepper is used so that it can be discernibly tasted in the dish.

This recipe is unusual for a shrimp Creole recipe. Glenda and I can quickly get enough of an acidic red gravy, so we used roux to tone down the red bite. The amount of bell pepper called for is also less than for most shrimp Creoles so as to reduce its dominance in the dish.

¼ cup flour
⅓ cup cooking oil
1 cup hot water
1 lb. peeled shrimp, cut in half if desired
1 8-oz. can tomato sauce
½ cup chopped green onion
½ cup chopped parsley
¼ cup chopped bell pepper
3 small cloves garlic, chopped
1¼ tsp. salt
½ tsp. thyme
⅛ tsp. cayenne pepper
1 bay leaf
1 lemon slice
Cooked rice

Make a roux by cooking flour and oil in a heavy pot. Stir constantly to keep from burning. When the roux turns dark brown, gradually add water and cook until thick and smooth, stirring constantly. Add remaining ingredients, except rice. Cover and simmer 20 minutes. Serve over rice. Serves 4.

Easy Roux

Making a roux may be the most difficult thing for someone new to Louisiana cooking to master. The constant stirring is tiresome, and it is best made in a cast-iron pot, which few cooks except throwbacks like us habitually use. Worse, making a nice dark roux involves nearly burning the flour. Some cooks simply avoid making roux by buying pint jars of pre-made roux sold in grocery stores.

Glenda quit making roux on top of the stove long ago when she lost a fight with it. While she was making a roux, one of our children startled her; the wooden spoon flipped out of her hand, catapulting a gob of near-done roux (this stuff is incredibly hot) onto the top of one bare foot. Instinctively, she wiped the top of her foot across the calf of the other leg, producing two second-degree burns instead of one.

She now uses—and has me using—a microwave to make roux. Ordinarily, I am not a lover of microwaves, considering them fit only to heat up leftovers and freshen bread, but this really works. It's easy and more economical than buying pre-made roux.

Simply put equal amounts of cooking oil and flour in a microwave-safe dish. Mix well with a whisk, cover with a paper plate, and cook on high in increments of a minute or less. Inspect and stir between each cooking. Shorten the cooking times as the roux nears the desired color. Store what is left (after cooling) in a screw-cap jar in the refrigerator, just as you would the store-bought stuff.

Lime-Garlic Broiled Shrimp

Broiling is an increasingly popular method of seafood preparation. It is a healthy way of cooking and can taste very good.

2 lb. large shrimp, peeled
6 cloves garlic, minced
1 cup margarine
4 tsp. lime juice
1 tsp. salt
½ tsp. black pepper
1 cup chopped fresh cilantro

Rinse shrimp and arrange in a single layer in a 10 x 15-inch baking dish. In a small saucepan, sauté garlic in margarine until tender. Remove from heat and stir in lime juice, salt, and pepper. Pour sauce evenly over shrimp. Broil for 8-10 minutes or until the shrimp are pink and tender. Mix cilantro in with shrimp and sauce near the end of cooking. Serves 4-6.

Tip: Be careful not to overcook this dish, though, because shrimp are even more likely than fish to dry out if cooked too long. Basting several times during cooking is helpful, but above all get the shrimp out of the oven when they are done.

Photo by Chris Granger

Shrimp Vanney

When we tasted this dish, it brought back a flood of memories from when Glenda and I moved to New Orleans nearly 40 years ago. Italian breadcrumbs laced with garlic tied together New Orleans dishes ranging from stuffed artichoke to mirliton and seafood casseroles. If you love old-time New Orleans cooking, you will love this one.

The recipe we started with came from the book *An Island Between the Chef & Rigolets*, an accounting of life in the Lake Catherine area in the 1900s by Arriollia "Bonnie" Vanney. I did make one significant change in Bonnie's recipe. The original called for 4 tablespoons of minced garlic instead of the 2 that I use.

Bonnie told me that she and her husband, Val, dreamed up the idea for this recipe while they were de-heading a couple of hundred pounds of shrimp and complaining about eating the same dishes repeatedly. After some kitchen engineering, this is what they came up with. Bonnie and Val were Lake Catherine commercial fishermen who fished almost exclusively for direct sale to the public. She was Val's deck hand on the boat before they became a couple and married. Bonnie, who now lives in Slidell, Louisiana, has two books about the Lake Catherine community. Her books can be ordered from arrivanney@bellsouth.net or 985-643-8177.

1 lb. very large shrimp
1 lb. sliced bacon
1 medium onion, minced
6 green onions, minced
2 tbsp. minced garlic
3 tbsp. chopped parsley
½ cup Romano cheese
1 tbsp. salt
1 tbsp. pepper
1½ cups Italian breadcrumbs
7 tbsp. olive oil
¼ lb. butter
¾ cup lemon juice

Peel and deeply butterfly each shrimp. Slice the bacon strips in half. Wrap the bacon around the shrimp and secure with a toothpick. Place each wrapped shrimp tail end up in a baking dish. Repeat until all the shrimp are prepared.

In a small bowl, combine onion, green onion, garlic, parsley, cheese, salt, pepper, breadcrumbs, and olive oil. Mix until the stuffing clings together. Place a teaspoonful of stuffing on each shrimp. Melt the butter in a small pan. Add the lemon juice to the melted butter and spoon over the top of the shrimp. Bake the shrimp uncovered in a 350-degree oven until done, about 15 minutes. Serves 4.

How to Peel and Butterfly Shrimp

Break the head from the shrimp and discard.

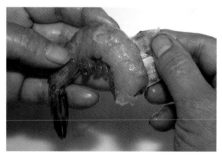

Grasp two of the segments of the shell along the bottom of the shrimp and peel upwards and around. Discard the shell. Repeat for the remainder of the segments.

Pinch the tail segment and pull the body away from it. Discard the tail.

Shallowly slice down the back of the shrimp over the length of the body. Remove the vein from the slit and discard.

To butterfly the shrimp, cut the deveining slit deeper. Spread the two sides of the shrimp open like butterfly wings

Baked Shrimp

This delicious dish really lets the taste of shrimp shine. It is simple and quick to prepare too.

1 stick margarine, divided
2 lb. medium to large shrimp tails
1½ tsp. salt
½ tsp. black pepper
¼ tsp. red pepper
4 tsp. flour
¼ tsp. ground nutmeg
2 cups half-and-half
2 egg yolks, beaten
4 tbsp. white wine
1 cup grated colby and Monterey Jack cheeses
1½ cups soft breadcrumbs
Paprika

Coat the bottom and sides of an 9 x 13-inch glass baking dish with 1 tbsp. of margarine. Spread shrimp one layer deep in the dish and sprinkle salt, black pepper, and red pepper evenly over them. In a medium saucepan, melt the rest of the margarine. Add the flour and nutmeg and stir over medium heat to blend. Gradually whisk in half-and-half. After the sauce has thickened, add beaten egg yolks and whisk. Mix in the wine and heat a minute longer. Spread mixture evenly over the shrimp. Sprinkle cheese over the shrimp and top with breadcrumbs. Sprinkle with paprika and bake at 350 degrees for 20-25 minutes. Serves 4.

Tip: Three slices of bread will provide about 1½ cups of breadcrumbs. The soft breadcrumbs will toast on the dish delightfully. Don't leave out the nutmeg.

Using Nutmeg

More so than some spices, ground nutmeg deteriorates quickly, so be sure it is fresh. Better yet, buy whole nutmegs, which keep indefinitely, and grate them as needed. Nutmeg is usually considered a spice for sweet dishes, but it works great with cheese and potato dishes as well as with some soups and sauces.

Whole nutmegs are about an inch long and are produced by large tropical evergreen trees. It is actually a fruit seed kernel and not a nut, so it does not pose a risk for people who have nut allergies.

Mace, produced from the same tree, is the dried lacy covering over the nutmeg. It has a more delicate flavor and a lighter color than its counterpart.

Shrimp Conga

Everyone loves shrimp; however, not everyone likes blue cheese. It's definitely an acquired taste, but once acquired, it becomes a passion. In this recipe the blue cheese does not overwhelm the shrimp, but you can definitely taste its pungent flavor.

2 sticks butter
1 cup cream cheese
⅓ cup crumbled blue cheese
2 lb. medium to large peeled shrimp tails
½ cup lime juice
Salt and pepper to taste

Melt butter in a small saucepan. Add cream cheese and blue cheese and mix until softened. Arrange shrimp in a large, deep baking dish. Sprinkle with lime juice, salt, and pepper. Spoon the cheese mixture over the top. Bake uncovered in a 400-degree oven until shrimp are pink and done. Be sure to serve with warm French bread to sop up the butter-cheese sauce. Serves 4-6.

Photo by Chris Granger

A Blue Cheese Primer

Blue cheeses, also called bleu cheeses, come in an enormous variety ranging from incredibly mediocre to indescribably scrumptious. One thing that they all have in common is the blue or blue-green veins or spots throughout. The color is due to molds (*yuck*) of the genus *Penicillium*, closely related to the mold that produces the antibiotic penicillin. Without the mold the cheeses themselves would be rather mundane.

Cheese makers introduce the mold by using long skewers (as with Roquefort) or by mixing the mold with the curds before pressing (like with Gorgonzola). Some blue cheeses, such as Danablue or Danish blue cheese, are of more recent origins and were developed to be less pricey competitors to expensive Roquefort, Gorgonzola, and Stilton cheeses. America's Maytag blue cheese, made from pasteurized milk, was developed at Iowa State University in 1941 and produced by Fred Maytag II of washing machine fame.

Usually blue cheese is made from cow's milk, but some are made from sheep's or goat's milk. After the mold is added, the product is aged in caves, often where the species of *Penicillium* used in their making was discovered. During aging, the mold grows inside the cheese and produces changes in the chemistry of the cheese that alter its physical structure and produce that unique tang and smell. Because of the range in quality of blue cheese, it is not advisable to purchase cheap pseudo varieties.

Jacked Shrimp

This recipe comes from Jack Rauenhorst of Sumrall, Mississippi. We had eaten bacon-wrapped, breaded, and fried shrimp before trying this recipe, but not with cheese in them.

1½ lb. very large shrimp tails
Old Bay Seasoning
Creole seasoning
Salt and white pepper
1½ lb. pepper jack cheese, sliced
½ lb. bacon slices, halved
2 eggs
1 cup heavy cream
1 cup flour
Cooking oil

Peel, devein, and butterfly each shrimp. Sprinkle with Old Bay and Creole seasonings, then salt and pepper to taste. Place a slice of cheese in the split of each butterflied shrimp. Pinch the shrimp closed and wrap each shrimp in half a slice of bacon. Secure with a toothpick. Whip eggs and cream together. To the flour add salt and pepper to taste. Dip each shrimp in the egg wash and dredge in the flour. Repeat the process, again dipping the shrimp in the egg wash and dredging in the flour. Fry in cooking oil until golden brown. Remove toothpicks before serving. Serves 4.

Tip: It is important to use the largest shrimp possible for this very good dish.

Are Bigger Shrimp Better?

Large shrimp cost a lot more than small shrimp. This has nothing to do with taste, but everything to do with scarcity. There are fewer big shrimp in the sea because human and fish predators catch them before they can grow large.

Interestingly, Louisiana commercial shrimpers, folks who know good shrimp, almost universally prefer medium-sized shrimp, 31-40 count per pound or smaller. They only use very large shrimp for preparations that involve a lot of manipulation, like the preceding one, or with harsh cooking styles that can dry shrimp out, such as grilling. Most of the time, shrimpers cite the tenderness of medium-sized shrimp as the reason for their preference.

From William D. Chauvin Collection, Louisiana State Archives, Baton Rouge

Fried Shrimp à l'Anglaise

One of the wonderful things about shrimp is that they can be so good with so little done to them. In fact, the less they are spiced, beyond basic salt and pepper, the better they may be.

2 lb. headless medium to large shrimp, peeled
2 eggs
1 cup milk
1 cup flour
Salt and pepper to taste

Rinse the shrimp and drain them. In a small bowl, beat eggs slightly. Add milk and mix. In another bowl, mix flour with salt and pepper. Dip shrimp in egg and milk wash, then roll them in flour. Heat the oil in a large skillet or deep fat fryer until hot, but not smoking. Drop the shrimp, a few at a time, into the hot oil and fry over medium heat until done to a light golden color, about 3-5 minutes. Serves 4-6.

Tip: Glenda always prefers to use wheat flour when frying shrimp and soft-shell crabs, and corn flour (fish fry) for fish and oysters. For those concerned with the unhealthy reputation of frying, use canola or peanut oil. They are heart-healthy like olive oil but have a higher smoking point, the temperature at which they begin to burn. Now if we could just get those calories out . . .

From William D. Chauvin Collection, Louisiana State Archives, Baton Rouge

French-Style Fried Shrimp Heads

Thanks to Ethel Guidry of Lafitte, Louisiana, for this unusual and really super recipe. When most people de-head shrimp, the heads go in the garbage and the tails in the frying pan. This recipe uses the heads. Cleaning the shrimp heads was a snap after her husband, Clinton (or Uncle Pops, as he was usually called), showed me how to do it. Simply pinch each head near the eyes and the meat pops out of the shell. This is a recipe that is well known to shrimpers but virtually unknown to the average seafood consumer. The finished product is occasionally referred to as "shrimp spiders" because the crispy legs give each head a spidery look.

½ cup milk
4 eggs
Fish fry
Salt and pepper
2 lb. shrimp head meats
Cooking oil

Make a wash by mixing together milk and eggs. Season the fish fry with salt and pepper to taste. Pass the meats through the wash and then through the fish fry. Using a strainer to hold the meats, deep fat fry for 7-9 minutes. These can be served as a meal or used as a tasty snack. Serves 4-6 as a meal and a lot more as an appetizer.

Preparing Shrimp Heads

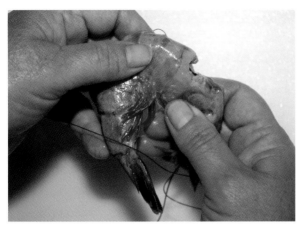

Break the head from the shrimp and set the shrimp aside.

Pinch the head near the eyes with one hand and gently pull on the shrimp's legs with the other hand.

After rinsing, the shrimp head is ready to bread and fry.

Is the Oil Ready?

Producing a good fried product depends largely on knowing when the cooking oil is hot enough to begin frying. If the oil is not hot enough, shrimp will become a soggy mess. If the oil is too hot, the shrimp will be dark brown and taste burned instead of golden with a crunchy crust. Most frying is done between 325 degrees and 375 degrees.

The most accurate method of checking the temperature of oil is with a thermometer, but not everyone has one and it can be hard to use in the shallow pan used for pan-frying. An old tried and true method of determining if the oil is ready is to drop a small square of bread in the oil. If it rises to the surface sizzling and frying, the oil should be ready to cook. If the bread takes 60 seconds to brown all over, the oil is 350-365 degrees. If it takes 40 seconds, it is 365-380 degrees. If it browns in 20 seconds, the oil is 380-390 degrees.

Boiled Shrimp

Since seafood boils are communal affairs in Louisiana, the proportions of this recipe are larger than for the others in this book. A very safe rule of thumb is to buy 2 lb. of head-on shrimp or 1 lb. of headless shrimp per person for boiling. Even if serving the heartiest of eaters, you will not need more than this if you cook the accompaniments.

20 lb. head-on shrimp
4 lb. small red potatoes
3 lb. onions
6 lemons
2 lb. smoked sausage
3 bags crab boil
1 8-oz. jar liquid crab boil
4 oz. cayenne pepper
10 ears of corn on the cob, halved
65 oz. salt
2 8-lb. bags of ice

Wash the shrimp and remove any small crabs or fish. Wash the potatoes, cut the onions in half vertically from stem to base, and cut the lemons in half crosswise. Cut the sausage links into 3-inch pieces. Fill a 60-qt. pot half-full of water. Light the fire on the seafood boiler and begin to heat the water. Add the potatoes, onions, lemons, bagged crab boil, liquid crab boil, and cayenne pepper. Cover the pot and bring to a boil. As the water nears boiling, taste the water by catching a few drips from a spoon on a finger. It should taste almost too spicy. If it doesn't, add more pepper and crab boil. Bring the water to a rolling boil and boil for 10 minutes. Add the shrimp, corn, and sausage. Make sure all the ingredients are covered by water. Return to a boil. If the shrimp are small, turn the fire off. If they are medium-sized, boil for 2 minutes. If they are large, boil for 5 minutes. Turn heat off and add the salt. Stir the water with a paddle or by carefully and slowly sloshing the pot liner up and down. Add the ice and when it has nearly melted, mix the water again. Allow the shrimp to soak in

Tip: Additional accompaniments can be added. If you plan to include celery stalks or carrots, cut the vegetables in halves or thirds and add them at the same time as the potatoes in the recipe. If mushrooms, asparagus, and wieners are desired, they should be added with the shrimp.

the water until the seasoning in the shrimp is strong enough for your taste. Sample the shrimp periodically. When the seasoning is right, remove the shrimp and drain. Serves 10.

Preventing Sticking Shells in Shrimp

Shrimp are one of the easiest types of seafood to cook, unless one overcooks them. One symptom of overcooked shrimp is shells that stick tenaciously to the meat in certain preparations such as barbequed shrimp or most especially, boiled shrimp. Sticking shells can ruin a shrimp boil.

The type of shrimp used can make a difference. White shrimp and sea bob shells are less prone to sticking than those of brown or pink shrimp. Sea bobs are usually considered too small for barbequing, so white shrimp should always be used for this dish and its many variations.

Using only white shrimp and sea bobs will not solve all sticking shell problems with boiled shrimp. I have seen or tried virtually every slick trick used to prevent sticking, including adding butter or olive oil to the water, timing the boiling with a stop watch, and crashing the post-boiling water temperature with loads of ice. Nothing works consistently, except for one thing.

The solution is to reserve the salt until the shrimp are done boiling. Do not add the salt along with the other seasonings prior to the boil. Refraining from adding the salt before the boil affects the taste very little because shrimp are boiled for such a short amount of time. Since shrimp, like other boiled seafood, are always soaked in their boiling water, they will pick up the salt with the other seasonings during soaking.

Lowcountry Boil

Courtesy Louisiana Seafood Promotion and Marketing Board

Convincing Louisianians that other people can boil or steam seafood as well as they do is a difficult task. In fact, if you pass a positive comment on boiled or steamed seafood from somewhere else, a Louisiana native will usually look at you as if you are crazy. But even Louisianans should like this dish, also known as Frogmore stew. It is South Carolina's unofficial seafood dish. The name comes from the community of Frogmore on St. Helena Island, South Carolina. The dish was created in the 1950s by Richard Gay, who first cooked it for his fellow National Guardsmen on National Guard Day. As with Louisiana boiled seafood, there are many different recipes for Lowcountry boils, but we enjoy this one.

5 lb. smoked hot sausage, cut into 1-inch pieces
2 gal. water
1 6-oz. can Old Bay Seasoning
5 lb. halved new potatoes
2 lb. peeled and quartered small onions
8 cloves garlic, minced
2 lb. whole carrots
½ cup cider vinegar
15 ears of corn, cut into 2-inch pieces
5 lb. very large headless shrimp
Salt to taste
Cocktail sauce
Lemon wedges

In a large frying pan, brown the sausage and set aside with any drippings. In a very large pot, add water and Old Bay Seasoning. Add potatoes, onions, garlic, and carrots. Boil for 20 minutes. Add the sausage with drippings and the vinegar. Cook another 15 minutes. Add corn and shrimp. Cook an additional 4-5 minutes or until the shrimp turn pink. Taste and add salt if needed. Drain and spread on newspaper. Serve with cocktail sauce and lemon wedges. Serves 10.

Creole Shrimp Butter

This is a spread for hot rolls, bagels, or toast. Use it like you would plain margarine or butter, and it will add a nice shrimp taste. It is also very good on baked potatoes.

½ cup margarine or butter
½ lb. peeled cooked shrimp
2 tbsp. fresh lemon juice
2 tsp. prepared horseradish
¼ tsp. salt
¼ tsp. black pepper
Dash of cayenne
Paprika

Allow margarine or butter to soften at room temperature. Chop shrimp finely with a food processor or knife. Mix all ingredients thoroughly. Sprinkle with paprika for color. Store in the refrigerator. Makes 1½ cups of spread.

Tip: The horseradish in the dish is important for its pungency. Be sure it is fresh. Fresh horseradish should be white to creamy beige in the jar and can be kept up to 6 months in the refrigerator. If it is turning brown, throw it out because it is losing its zing and developing a stale taste.

A Taste of Horseradish

Horseradish gets its bite and aroma when enzymes released during grating break down sinigrin, the plant's natural pest deterrent, to produce allyl isothiocyanate (mustard oil). This pungent oil irritates a person's sinuses and eyes rather than sparking taste buds on the tongue. If a similar taste is noticed in wasabi by sushi and sashimi lovers, it is not surprising. Virtually all the "wasabi" sold in the U.S. and even in most sushi bars in Japan is imitation wasabi, made from horseradish, mustard seed, and green food coloring.

Cultivated for their roots, horseradish and true wasabi are both members of the *Brassica* family, which includes mustards and cabbages. True wasabi produces its taste with compounds very similar to horseradish. The big difference between the two is in ease of cultivation and price. Horseradish grows almost everywhere and can spread from cultivation to become an almost ineradicable weed. In contrast, true wasabi may be one of the most temperamental of cultivated plants, growing only on soggy mountain terraces watered by alpine streams that have a constant year-round temperature of 51-57°F. Few places are ideal for wasabi growing, hence the very high price for the real article.

Aloha's Barbecued Shrimp

Aloha Moll, formerly of Harvey, Louisiana, presently of Poplarville, Mississippi, first prepared this dish for us during the 1980 Christmas holidays. Usually dishes with this many seasonings run the danger of being "muddy" tasting. This one doesn't have that problem. Aloha admitted that she put this together from a dozen different recipes. Spicy but delicious, it is the barbequed shrimp dish that we most often make. Aloha's father-in-law, Jerry Moll, was a career shrimper from Jefferson Parish, Louisiana.

6 lb. shrimp tails
¼ cup olive oil
3 sticks butter
5 cloves garlic, minced
6 tsp. barbeque seasoning
2 tbsp. black pepper
1 tbsp. Worcestershire sauce
1 tsp. salt
1 tsp. chopped rosemary
1 tsp. dried oregano
1 tsp. Creole seasoning
1 tsp. cayenne pepper
1 tsp. Tabasco sauce
2 tsp. liquid crab boil
2 large onions, thinly sliced
2 stalks celery, coarsely chopped
3 lemons, sliced
3 loaves French bread

Photo by Chris Granger

Rinse and drain shrimp well then mix thoroughly with olive oil to prevent the shells from sticking. Let sit half an hour. Melt butter in a small saucepan and mix in all seasonings except onions, celery, lemons, and bread. Place shrimp in a 9 x 13-inch baking pan and pour mixture over shrimp. Mix well. Spread onions, celery, and lemons over top. Bake 35-45 minutes in oven at 350 degrees, basting shrimp with the sauce every 5-10 minutes. Serve with French bread to sop up the sauce. Serves 6 generously.

Liquid Crab Boil: The Magic Bullet

Liquid crab boil is a "magic bullet" ingredient and a small bottle should be in everyone's kitchen cabinet. Its value is not in making dishes peppery hot: I liken it to liquid Creole seasoning without all the salt. Among its ingredients are extracts of spices such as allspice, bay, mustard, thyme, and clove. In constructing a dish, we occasionally hit a taste dead-end, but a teaspoon of liquid crab boil will often bring that dish to life. Liquid crab boil is also a wonderful flavor builder in traditional dishes such as Cajun white beans.

Barbequed Shrimp Usannaz

We first tasted this recipe in the kitchen of May Usannaz's crab shop in the community of Venetian Isles in extreme eastern New Orleans. A crab and freshwater eel buyer, May had deep roots in the seafood industry, while her husband, Noel, prowled the Louisiana coastline in his 57-foot shrimp trawler, *Gros Comme' Ca,* which translates as "Big Like This."

Noel Usannaz himself was bigger than life. He was the consummate old-time professional fisherman. When I accidentally stepped on one white shrimp out of the thousands heaped on the deck after the trawls were emptied, I received a stern lecture about waste. He explained that these shrimp were destined to feed people and should be treated with respect and care. With his dark hair, eyes, and complexion, he resembled nothing so much as Teche Bossier played by Gilbert Roland in the 1953 movie classic *Thunder Bay.*

Visiting May's crab dock was always fun. Besides being an engaging conversationalist, she always had something good to eat on the stove. This is Louisiana barbequed shrimp at its most basic. Using French bread to sop up the sauce is integral to the dish.

Courtesy May Usannaz

2½ lb. large headless shrimp
3 sticks margarine
4 tbsp. lemon-pepper seasoning
2 tbsp. dried parsley
2 loaves French bread

Wash the shrimp. Melt margarine and mix in the lemon-pepper seasoning and parsley. Pour into a baking dish. Add shrimp and stir well to coat. Bake at 350 degrees, stirring several times during cooking, for 45 minutes or until shrimp are done and peel easily. The shrimp and margarine will make its own sauce. Serve with plenty of French bread to sop up the sauce. Serves 4.

Barbequed Shrimp

This variation of barbequed shrimp is from Chantel Ann Lincoln of St. Tammany Parish, Louisiana. Note the large amount of garlic in this excellent dish. Surprisingly, it doesn't overpower the shrimp.

4 lb. large head-on shrimp
1 lb. butter
¼ cup Worcestershire sauce
5-8 drops hot sauce or to taste
Black pepper to taste
10 cloves garlic, finely chopped
3 tbsp. chopped fresh rosemary
2 loaves French bread

Wash the shrimp in cold water, leaving the heads on. In a large Dutch oven or oval roaster, melt the butter on top of the stove. Add Worcestershire sauce (more Worcestershire sauce can be added later to taste) and hot sauce. Add the shrimp to the sauce and sprinkle with black pepper. Sprinkle garlic and rosemary over the shrimp and toss lightly. Be careful not to break the heads off of the shrimp. Remove from stove and place uncovered in a 350-degree oven for 35-40 minutes. Toss lightly once or twice during baking. Bake until the shrimp have separated slightly from their shells. Serve with French bread for dipping in the sauce. Serves 4.

Who Invented Barbequed Shrimp?

Seemingly a thousand variations of barbequed shrimp exist. Pascal's Manale Restaurant on Napoleon Avenue in New Orleans is almost always given credit for producing the original version. No one seems to know how the dish received its name as it has absolutely nothing to do with barbequing, even with today's stretched definitions. It's just another example of the wonderful originality of Louisiana's chefs and cooks and their penchant for whimsical names. Pascal's Manale was founded in 1913 and is most typically described as a Creole-Italian restaurant. Interestingly, while the restaurant is most famous for its barbecued shrimp, its pride is in its primo raw oysters.

Lemon-Soy Barbequed Shrimp

Tip: The larger the shrimp used, the less likely they are to become over-cooked. Overcooking is the only no-no with this dish. You will need skewers—metal, wood, or bamboo—to prepare this recipe.

I have nothing against what are called "barbequed" shrimp in New Orleans; in fact, they are delicious. But nowhere else would shrimp swimming in butter and black pepper and cooked in the oven pass for barbeque. This recipe is for real barbequed shrimp, the kind on a grill. And it is really good.

2 lb. large shrimp, peeled
2 cloves garlic, minced
½ tsp. salt
½ cup soy sauce
½ cup lemon juice
3 tbsp. finely chopped parsley
2 tbsp. minced onion
½ tsp. pepper

Rinse shrimp and set aside. Mash garlic with salt and place in a medium-sized bowl. Add remaining ingredients and mix. Add shrimp and marinate for 30 minutes. Thread shrimp on skewers and grill 5-10 minutes, turning and basting with remaining marinade twice. Do not overcook. Serves 4.

Using Soy Sauce

Soy sauces are used to bring out flavors in many types of food—meats, vegetables, and soups. Its combined sweet, sour, salty, and bitter qualities perk up flat-tasting foods. Soy extract has well-known flavor-enhancing properties known as "umami," a word borrowed from the Japanese language. Made from soybeans and roasted grain, soy sauces are high in antioxidants but are also high in salt, making their use problematic for those on low-sodium diets. Low-salt soy sauces do exist, but it is impossible to make soy sauce without some salt.

Shrimp Mosca

Tip: White shrimp are better for barbequing than brown shrimp, as they peel much easier.

This recipe comes from Charles Arcement of Lafitte, Louisiana. Charles is a commercial shrimper and knows good shrimp. In taste, this dish resembles New Orleans-style barbequed shrimp, but it does not have all the butter and oil in it.

8 lb. large head-on white shrimp
4 large onions, sliced
1 stalk celery, chopped
6 lemons, sliced
1 bunch green onions, chopped
2 garlic heads, peeled and sliced
1 lb. small potatoes
Salt and pepper to taste
1 5-oz. bottle Pickapeppa Sauce
1 5-oz. bottle Worcestershire sauce
3 12-oz. cans beer
2 loaves French bread

Place a layer of shrimp in the bottom of a large pot. Layer a portion of the onions, celery, lemons, green onions, garlic, and a few potatoes over the shrimp. Sprinkle with salt and pepper. Repeat layers until all layering ingredients have been used. Pour Pickapeppa, Worcestershire, and beer into the pot. Cook until shrimp are done, not more than 30 minutes. Serve with French bread to sop up the sauce. Serves 6.

The Roots of Shrimp Mosca

The many variations of Shrimp Mosca are mostly Italian-style barbequed shrimp. Shrimp Mosca has been credited both to Mosca's Restaurant, located in the swamp off Louisiana Highway 90 between Avondale and Westwego, and La Louisiane Bar & Bistro in the New Orleans French Quarter. Interestingly, the name "Shrimp Mosca" appears nowhere on Mosca's never-changing menu. Many variations of this recipe exist, and most have the shrimp swimming in olive oil and include white wine, typically sauterne. This one uses beer and includes Pickapeppa Sauce, the somewhat sweet condiment often called "Jamaican ketchup." Other variations of this dish may include Italian seasoning, Italian breadcrumbs, Italian cheese, or bay leaves.

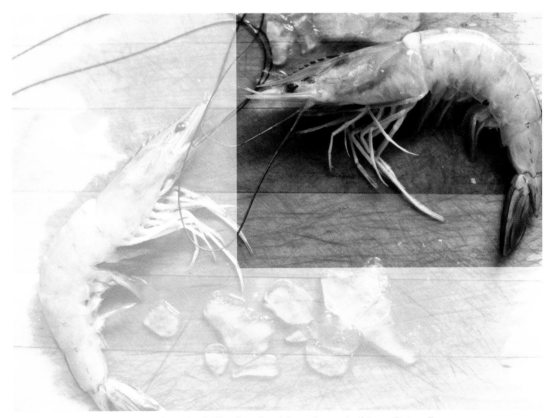

From William D. Chauvin Collection, Louisiana State Archives, Baton Rouge

Shrimp Brenda

Brenda Tullos used to cook lunch for her husband Frank and herself at their retail seafood shop, Frank's Riverside Seafood, in Metairie, Louisiana, which is where I first sampled this dish. Of course, Brenda's menu more often than not featured seafood.

Many Louisiana shrimp dishes revolve around olive oil and/or butter and garlic. While garlic is not a member of the "holy trinity" of onions, bell pepper, and celery in Cajun and Creole cooking, it is heavily used, especially in the Creole cooking of the city of New Orleans and surrounding area.

2 lb. medium to large headless shrimp
1 stick butter
2 tbsp. olive oil
6 cloves garlic, minced
1½ lb. sliced mushrooms
1 bunch green onions, diced
Paprika
Salt and pepper to taste
2 loaves French bread

Peel and wash shrimp. In a large saucepan, melt butter, add oil, and sauté garlic until soft. Add shrimp and cover. Cook over medium-low heat until done, 12-15 minutes. Stir in mushrooms and green onions and sprinkle paprika for color. Season with salt and pepper to taste. Cook 5-10 minutes until mushrooms are tender. Serve with hot French bread. Serves 4.

The Stinking Rose

Garlic has been called "the stinking rose" since at least as early as the Roman Empire, though the origin of the name is forever lost. It is the most pungent member of the onion family, which includes shallots, leeks, and, of course, onions. In spite of humankind's taste for garlic, its powerful odor and taste likely evolved to prevent insects and other animals from eating it.

The compound diallyl disulfide is responsible for its odor and allicin, a clear liquid compound, for its spicy taste. Both are sulfur components and both are released as garlic cells are cut during processing and their fluids escape. Cooking garlic mellows its pungency somewhat by removing allicin, although a garlic smell may still occur in a diner's breath and sweat a day after eating garlic. This odor results because some of garlic's strong-smelling sulfur compounds are metabolized in the body into a form that is not digestible. Since the body is unable to break the compound down, it passes intact into the bloodstream, where it is carried to the lungs and skin and released.

Shrimp Kebabs

Tip: You will need a charcoal grill and shish kebab skewers for this recipe. If you have a rotisserie, so much the better. If you only have a gas grill, you are just as well off cooking this dish in your range's oven. Meats, like beef or pork, may develop a smoky taste on a gas grill because their fats drip onto hot surfaces and volatize as smoke. Shrimp are lean, as are the other ingredients in this dish, so no fat is produced and thus, no smoke. Also, use white shrimp for this dish, as the shells peel much easier than those of brown or pink shrimp after cooking.

This is something that I put together one weekend to impress Glenda and it came out really good.

1½ lb. large shrimp tails
1 pt. cherry tomatoes
½ lb. fresh mushrooms
1 bell pepper, cut into one-inch squares
1 14-oz. can pineapple chunks, drained
½ cup cooking oil
½ cup lemon juice
½ cup dried parsley flakes
½ cup soy sauce
1½ tsp. salt
1 tsp. black pepper

Rinse shrimp tails and drain well. Do not peel. Place shrimp, tomatoes, mushrooms, bell pepper, and pineapple in a bowl. Combine rest of ingredients, mix, and pour over the shrimp mixture. Marinate 30 minutes, stirring once or twice. Place alternately the shrimp, mushrooms, tomatoes, bell pepper, and pineapple on skewers. Cook over a moderately hot fire and baste with remaining sauce. When the cherry tomatoes become soft, no longer than 10-12 minutes, the shrimp are ready. The skewers may be removed before serving or the food may be served on the skewers. Serves 4.

Creole Shrimp

Allen Wiseman of Lafitte, Louisiana, who provided this recipe to us, is a lifelong commercial shrimper. He has cooked many a shrimp dish in one shrimp boat galley or another. This dish is similar to barbequed shrimp but has the spicy taste of Creole mustard.

3 lb. large headless shrimp
1¼ lb. margarine
1 16-oz. jar Creole mustard
1 tbsp. chopped fresh chives
1 tbsp. dried parsley flakes
1 tsp. red pepper
2 loaves French bread

Wash shrimp and set aside. Melt margarine in a large electric skillet or baking dish. Add the mustard, chives, parsley, and pepper. The mustard is well seasoned and usually adds enough salt for most people's taste. Mix the raw shrimp into the mixture. Cover and bake at 350 degrees for 20-30 minutes or until shrimp are done and the water is cooked off. If an electric skillet is used, set it at 325 degrees and open the small vent in the cover, cooking until no more water remains. After the shrimp are cooked, you may want to pour the excess oil off the top of the mixture. Serve with French bread to sop up the sauce. Serves 4-6.

Tip: Do not substitute dried chives for fresh chives. While most dried herbs are acceptable substitutes for fresh, two are not—rosemary and chives. Rosemary dries so hard that is virtually petrified and impossible to soften while chives are utterly tasteless after drying. If fresh chives are unavailable, use finely minced green onion tops instead. Fortunately, both rosemary and chives grow well year round in Louisiana. Any serious cook should consider growing at least these two herbs.

Drunk Shrimp

Tip: Various versions of drunk shrimp can be found along coastal Louisiana, almost invariably in commercial shrimpers' hands. Drunk shrimp is always prepared in large batches and is typically cooked and eaten outdoors.

We have to thank James Frickey and his son Jimmy of Westwego, Louisiana, for this recipe. They first prepared it for us in 1981. It is as good now as it was then. Jimmy and the late James both shrimped out of Grand Isle, Louisiana, where their recipe was legendary.

10 lb. large headless shrimp
6 lb. onions, chopped
1 stalk celery, chopped
3 bunches green onions, chopped
1 lb. butter
1 qt. olive oil
½ lemon, finely chopped
¼ cup ketchup
1 bell pepper, finely chopped
1 tbsp. Worcestershire sauce
2 jalapeno peppers, finely chopped
3 12-oz. cans beer
½ gal. sauterne wine
Salt to taste
½ tsp. liquid crab boil
1 tbsp. chopped parsley
1 head garlic, peeled and minced
French bread

Wash shrimp. In a large pot cook onions, celery, and green onions in butter and olive oil over medium heat until tender. Stir, being careful not to brown or burn the vegetables. Add lemon and cook 5 minutes. Separately stir in ketchup, bell pepper, Worcestershire sauce, and jalapeno, cooking each ingredient 5 minutes before adding the next. Add beer and wine, cover, and heat to boiling. Boil about one hour. If liquid runs low at any time in cooking process, add beer, not water. Season with salt and crab boil and cook 5 more minutes. Add parsley and garlic and cook 15 additional minutes. Add shrimp and cook for 10 minutes. Turn off heat and let soak for 30 minutes. Serve with French bread to sop up the sauce. Serves 12.

Other Drunk Shrimp

Louisiana does not have a monopoly on shrimp dishes by this name. A dish called drunk shrimp can often be found on Mexican or Central American restaurant menus as *camarones borrachos*. The alcohol used in Latin drunk shrimp is typically rum. The most bizarre drunk shrimp preparations take place in various Asian countries. In these preparations (if you can call them that), live shrimp are immersed in an alcoholic potion for a short period of time, then eaten while still alive, but in theory, drunk.

From William D. Chauvin Collection, Louisiana State Archives, Baton Rouge

Pancit Bihon

Shrimp are eaten the world 'round. This recipe is from Eden Tablan, a native of Bataan in the Philippines who now resides in Louisiana. Pancit bihon is the second most popular food in the Philippines, following only rice. The name translated literally means "something conveniently cooked fast." It is very good and resembles a south Louisiana-style dish.

Tip: The rice noodles and the oyster sauce are available at any Asian/ Vietnamese grocery store. The broths can be made by simmering pork scraps and shrimp heads in water in separate pots. Simmer the broth down until it has a nice color. If you just can't get around to making your own broths from scratch, store-bought pork or shellfish bouillon or base will work. If you can't find either, substitute chicken bouillon or base.

1 package rice noodles
4 cloves garlic, minced
4 tbsp. cooking oil, divided
1 medium onion, slivered
1 lb. peeled shrimp
¼ small head cabbage, shredded
3 carrots, diced
6 tbsp. oyster sauce, divided
1 cup pork broth, divided
1 cup shrimp broth, divided
Salt and pepper
3 green onions, chopped

Cook noodles according to package directions, then drain. While noodles are cooking, sauté garlic in 2 tbsp. of oil in a wok or large pot over medium heat until light brown. Add onion and sauté until clear, then add shrimp. Sauté until shrimp turn pink. Remove everything from wok and set aside. Heat remaining oil in the wok and stir-fry cabbage and carrots over high heat until tender. Lower heat to medium. Stir in 3 tbsp. oyster sauce. Remove vegetables from wok. Put drained noodles in wok. Pour ½ cup pork broth, ½ cup shrimp broth, and 3 tbsp. oyster sauce over noodles. Cook over low heat until noodles are soft, adding stock as needed until noodles are cooked. Return shrimp mixture and vegetables to the pot and heat until hot. Season with salt and pepper to taste. Sprinkle chopped green onions as garnish. Serves 4-6.

Gingered Shrimp with Pea Pods

The art of Chinese cooking has become quite popular in this country. This recipe involves stir-frying, a fast and easy technique. While this dish can be prepared in a saucepan, the use of a wok makes the job much easier.

1½ lb. shrimp tails
¼ cup soy sauce
3 tbsp. white wine
½ cup chicken broth
2 tsp. minced fresh ginger
2 tbsp. cornstarch
¼ cup vegetable oil, divided
2 6-oz. packages frozen pea pods, thawed
3 green onions, cut into 1-inch pieces
1 8-oz. can sliced water chestnuts

Peel the shrimp and split them in half lengthwise if they are large. Combine soy sauce, wine, chicken broth, ginger, and cornstarch in a bowl then set aside. Heat half the oil over high heat. Add the shrimp and stir-fry until they are pink and opaque. Remove and set aside. Heat remaining oil. Cook pea pods over high heat, stirring rapidly for 3-4 minutes. Remove from wok and set aside. Repeat with onions then water chestnuts, cooking each vegetable separately until they are hot and just soft. Return all ingredients to the wok, add the soy sauce mixture, stir, and cook until the sauce thickens slightly, about 2-3 minutes. Serve with rice. Serves 4.

Tip: The secret to stir-frying is to cut up all of the ingredients before you begin cooking then add them one at a time to the hot oil, all the time "stir-tossing" the food rapidly. Each ingredient is only lightly cooked and the hot oil seals in the color, flavor, and crispness of vegetables and the natural juices of meats and seafood.

Use fresh ginger root when possible. Dried ground ginger is a weak-kneed substitute that never comes close to matching the complexity of taste found in fresh ginger.

Choosing a Wok

Woks are available in many sizes and are made of many materials. A traditional 14-inch carbon steel wok is easily the best choice for the average kitchen, and it's the least expensive. Cast-iron woks hold temperature well but are slow to heat up and heavy to handle. Stainless steel conducts heat unevenly and is expensive. Aluminum woks do not retain heat well and are easily damaged. Nonstick surfaces are easily damaged and some may quickly deteriorate with the high heat necessary for stir-frying. Electric woks heat poorly, are inefficient, and tend to be an all-around disaster.

Woks may have a round bottom or a flat bottom. Round-bottomed woks funnel food to the cooking point best, but on electric ranges, flat-bottomed woks must be used. Traditional woks are equipped with two D-shaped metal handles mounted opposite each other on or near the lip of the wok. These become very hot and difficult to handle with the high heat of stir-frying. A better option may be to purchase a wok with a wooden or plastic stick handle. Woks with such handles are sometimes called "Peking pans" or "pau woks."

Curried Shrimp

This is a quick but delicious version of British curry. With this dish, just a half-hour after you begin cooking, you will be seated at the table and eating.

1 lb. unpeeled cooked shrimp
¼ cup chopped onion
3 tbsp. margarine
3 tbsp. flour
1 tsp. salt
Dash of pepper
1 tsp. curry powder
¼ tsp. powdered ginger
2 cups milk
Cooked rice

Peel shrimp and if they are large, cut in half. Sauté onion in margarine until tender, then blend in flour and seasonings. Add milk gradually and cook until thick, stirring constantly. Add shrimp and heat well. Serve with rice. Serves 4.

The Many Types of Curry

Curries can be divided into two major groups, Indian and Thai, and two minor groups, British and Japanese. Indian curries are yellow in color and their preparation almost always starts with toasting ground spices in a pan. Indian curry spice mixtures contain curry tree leaves, turmeric (providing the yellow color), coriander, ginger, garlic, chili, black pepper, and tamarind. Some mixtures are comprised of up to 20 spices. Indian curries are usually named after their main ingredient, as with chicken curry or potato curry. Depending on the region of origin, an Indian curry will use milk or cream, clarified butter (ghee), or coconut milk.

Thai curries are quite different from Indian curries. They use fewer dry spices and more fresh herbs and garlic. Coconut milk is also very common in Thai curries and curry tree leaves are never used. The three major Thai curries are identified by color—green, red, and yellow. All are typically prepared from pastes. Green curries are based around young green chilies and are usually hotter than the others, which are often slightly sweet and contain cumin and coriander. Red curries use larger mature red chilies and are slightly milder than green curries. They often contain lemon grass, garlic, shallots, and Thai ginger (galangal). Yellow curry has the boldest, most aromatic taste of the three and is the most like Indian curry. Its yellow color comes from turmeric.

Three other Thai curry types are Massaman, Panang, and Prik Khing. Massaman curry is a mildly hot red curry. It is, by far, the sweetest of Thai curries, with a tart tang provided by tamarind. Chicken and shrimp are often used in this curry. Panang curry is also similar to red curry but is milder and sweeter. Beef often appears in Panang curry. Prik Khing curry is the least common and leans heavily on the use of garlic as well as other spices.

British curries were adapted from Indian curries during the long period of British rule over India. British curry is typically cooked to a gravy-like consistency. The pre-blended curry powders sold in America are British, though every curry powder brand is different because such a wide variety of spices can be chosen.

Curry is also popular in Japan but is widely viewed as a nontraditional western dish. Japanese curries are almost stew-like, containing potatoes, carrots, and onions, and have a thick dark brown gravy.

Shrimp Curry

This is an authentic Indian recipe that you will like if you are partial to curry. Instead of curry powder it uses cumin, turmeric, and coriander, which is one of the prominent ingredients in bags of crab boil. You may adjust the ratio of the three to get the taste you want.

1½ lb. peeled shrimp
¼ cup vegetable oil
1 medium onion, thinly sliced
1 1½-inch chunk fresh ginger root, peeled and pureed
4 large cloves garlic, pureed
2 tsp. ground cumin
2 tsp. ground coriander
1 tsp. ground turmeric
½ tsp. salt, plus more to taste
½ cup plain yogurt
1 cup + 2 tsp. chopped fresh cilantro leaves, divided
2 cups water
1 jalapeno pepper, cut in half
1½ cups thawed frozen green peas
1 tbsp. cornstarch, optional
¼ cup water, optional
Cooked rice

Rinse the shrimp and set aside to drain. Heat the oil in a large, deep skillet or an iron pot over medium-high heat until hot but not smoking. Add onion and sauté until softened. Stir in ginger, garlic, ground spices, ½ tsp. salt, and yogurt. Stir frequently until the liquid evaporates, the oil separates and turns orange, and you can smell the spices frying. Stir in 1 cup cilantro, water, and jalapeno. Reduce heat, cover, and simmer for 20 minutes. Add remaining cilantro, peas, and shrimp. Salt to taste. Simmer uncovered until shrimp are done. If curry is watery, mix together cornstarch and ¼ cup water and add slowly to sauce. Stirring, heat until thickened. Serve over rice. Serves 4.

Tip: If you can't find ground coriander, use the whole seed, crushed into smaller particles.

Jhinga Caldeen

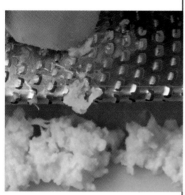

Tip: We used two small serrano chili peppers, but if you like your food really spicy, you may add another one. For a touch of authenticity, substitute basmati rice for white rice.

Shrimp curries are made all over Asia, and the variations from country to country or even province to province are endless. This dish is from Goa, the smallest state in India. Goa, located on India's west coast, was controlled by Portuguese merchants for about 450 years, until India retook control in 1961. Besides jhinga caldeen, Goa is famous for its beautiful beaches.

1½ lb. medium shrimp tails, peeled
1 oz. white wine
Salt
2 tsp. ground coriander
1 tsp. ground turmeric
1 tsp. ground cumin
¼ tsp. chili powder
½ tsp. black pepper
1 tbsp. vegetable oil
1 small red onion, chopped
1 1-inch piece of fresh ginger, grated
4 cloves garlic, minced
3 oz. water
2 oz. coconut milk
2 fresh chili peppers, seeded and finely slivered
½ cup chopped fresh cilantro
Cooked rice

Mix together shrimp, wine, and salt. Set aside 30 minutes to marinate. Mix coriander, turmeric, cumin, chili powder, and black pepper in a small bowl and set aside. In a large frying pan, heat oil and onion over medium heat until onion is caramelized (about 5 minutes). Add ginger and garlic and saute 1 minute. Add mixed spices and cook 2 minutes, stirring constantly. Add water, a dash of salt, coconut milk, and chilies. Simmer until coconut milk blends into the sauce, 1-2 minutes. Add marinated shrimp and cilantro and cook about 5 minutes or until done. Serve over rice. Serves 4.

Index